排云殿佛香阁正立面
Front Elevation of the *Paiyun* Hall and *Foxiang* Tower

霁清轩组群北立面
North Elevation of the *Jiqing* Hall Complex

『十二五』国家重点图书出版规划项目

中国古建筑测绘大系·园林建筑

颐和园（第二版）

天津大学建筑学院　北京市颐和园管理处　编写

王其亨　主编　张龙　张凤梧　编著

中国建筑工业出版社

Traditional Chinese Architecture Surveying and
Mapping Series:
Garden Architecture

SUMMER PALACE (2nd Edition)

Compiled by School of Architecture, Tianjin University &
Administration Office of the Summer Palace, Beijing
Chief Edited by WANG Qiheng
Edited by ZHANG Long，ZHANG Fengwu

China Architecture & Building Press

Contents

目　录

Location of Summer Palace: Western suburb of Beijing

The beginning of the construction: 1750

Occupied area: 300.8 hectares

Governing body: Administration Office of the Summer Palace, Beijing

Surveying and mapping unit: School of Architecture, Tianjin University

Time of surveying and mapping: 1956—2013

地　　址　北京西郊

始建年代　1750 年

占地面积　300.8 公顷

主管单位　北京市颐和园管理处

测绘单位　天津大学建筑学院

测绘时间　1956—2013 年

Introduction

Summer Palace occupies a site covering 300.8 hectares located in the northwest of the city of Beijing (Fig.1), 15 kilometers north west of the Forbidden City. The site includes Longevity Hill, *Kunming* Lake, West Dyke and other islands, forming picturesque scenery of 'hills encircled by water', 'dyke and islands reflected in lakes.' Summer Palace (formerly *Qingyiyuan*) was constructed in 1750, the 15th year of the Qianlong Emperor's reign; ruined in 1860, the 10th year of the Xianfeng Emperor's reign; reconstructed in 1886, the 12th year of the Guangxu Emperor's reign, renamed Summer Palace in 1888, listed as a key national conservation site of cultural heritage in 1961, and inscribed in the World Heritage List in 1998. There are over 3500 architectural artefacts, covering almost 70,000 square meters, in Summer Palace.

导　言

颐和园位于北京西北郊（图一），距故宫紫禁城约 15 公里，占地面积 300.8 公顷，由万寿山、昆明湖、西堤及诸岛屿构成了山水环绕、堤岛映带的景观格局。颐和园前身清漪园，始建于乾隆十五年（1750 年），咸丰十年（1860 年）被毁，光绪十二年（1886 年）重修，十四年（1888 年）更名为『颐和园』，1961 年颐和园被列为第一批全国重点文物保护单位，1998 年入选世界文化遗产名录。全园现存古建筑 3500 余间，总建筑面积近 70000 平方米。

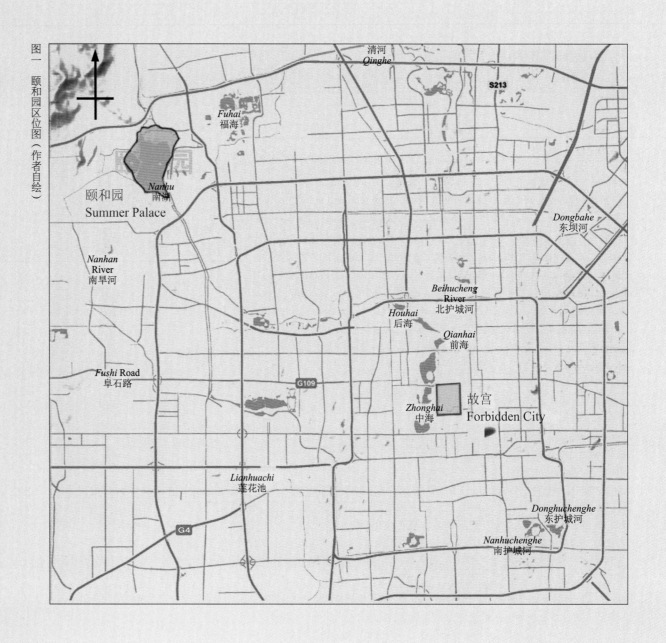

图一 颐和园区位图（作者自绘）

Fig.1　The Location of Summer Palace (Drawn by author)

The Construction of *Qingyiyuan*

1. Topography

Longevity Hill of Summer Palace, originally named *Weng* Hill or *Wengshan*, is one of the West Hills in Beijing. *Kunming* Lake (the former *Weng* Hill Lake or *Wengshanpo*), also named West Lake, is a low-lying area between the alluvial fan of the *Yongding* River and the alluvial fan at the base of *Nankou* Hill. According to archaeological research findings, this water area formed and stabilized 3500 years ago. The Jin Dynasty (1115—1234) established their capital in Beijing, then the Yuan Dynasty (1271—1368) further established there the *Yuandadu* (the greatest of all capitals). In order to provide the capital with an assured water supply and ensure access by water transport, Guo Shoujing built the *Baifu Wengshan* River; tapped springs on the river's way to join *Wengshanpo* Lake; supplied water transport through *Chang* River (*Changhe*), *Taiye* Lake and *Tonghui* River and then changed *Weng* Hill Lake from a natural into an artificial reservoir (Fig.2). The water level of the reservoir was controlled and its construction greatly promoted cultural activities around *Weng* Hill Lake.

"The fragrance floats like waves in the distance, the cloud is like fog, the rainbow strides over the west lake, the green pavilions hang in front of the sun and moon"; "The water is so blue when the sun sets in the lake in Spring; The pavilions and sky are reflexed in the water; You seem to walk as if in a drawing when you are in the green mountains; Two flying white birds remind one of the scenery of *Jiangnan* [the Southern part of the *Yangzi* delta]".

These poems, written during the Yuan and Ming Dynasties, describe the picturesque landscape of this area combining cultural and natural scenery as a water village.

During the late Ming and early Qing periods, Chinese society was in chaos and there was continuous war. So the West Hills lost their splendour and increasingly became ruins. After the Kangxi Emperor restored peace to the *Sanfan* area, Qing society and economy stabilized. Kangxi began to construct imperial gardens amidst the foundations of the palaces and private gardens of past dynasties in the west of Beijing. From the 16th year of the reign of Kangxi to the 9th year of the reign of Qianlong (1677—1744), *Jingyi* Garden in *Xiangshan* (Fragrance Mountain), *Changchun* Garden, *Jingming*

清漪园的营建

一、地形地貌

颐和园万寿山原名瓮山，是北京西山余脉。昆明湖前身瓮山泊，又称『西湖』，是永定河冲积扇和南口山冲积扇之间的低洼地带，据考古研究发现，这一水体早在3500年前就已经形成，并逐渐趋于稳定○。随着金代（1115—1234年）定都北京，尤其是元代（1271—1368年）元大都的兴建，为保证皇城、漕运用水，郭守敬开凿白浮瓮山河，将沿途诸泉汇于瓮山泊，再经长河、太液池、护城河、通惠河补给运河漕运用水，使瓮山泊从天然水库成为人工水库（图二），湖水水位得到控制，极大地推动了瓮山泊一带的人文开发。

『十里香风荷盖浪，一川云景柳丝烟。玉虹遥亘西湖上，翠阁双悬日月前。』○

『春湖落日水拖蓝，天影楼台上下涵。十里青山行画里，双飞白鸟似江南。』○

所描述的就是元明时期这一带人文与自然景观交织的水乡图画。

明末清初，社会动乱，战事频仍，西山一带也失去往日繁华，趋于破败。康熙平定三藩后，社会稳定、经济渐渐繁荣，开始在北京西郊历代行宫或私家园林的基础上建设皇家园林，自康熙十六年至乾隆九年（1677—1744年）先后修建了香山静宜园、畅春园、玉泉山静明园、圆明园四座皇

Fig.2 River System of the Western Outskirts of Beijing in Yuan Dynasty (From *Conservation Plan of the Summer Palace* made by Research Institute of Architecture Design of Tianjin University)

Fig.3 Relics of Foundations of Ancient Dyke Poles in *Kunming* Lake (Supplied by the Administrative office of the Summer Palace, Beijing)

Fig.4 *Weng* Hill and West Lake in the Drawing of Beijing–Hangzhou Transport (courtesy of the Museum of Zhejiang Province)

Fig.5 Diagram of the *Weng* hill Area before *Qingyiyuan* (ZHANG Long. *Comprehensive Research on Yangshi Lei Archives of the Summer Palace*[D]. Tianjin: Tianjin University, 2009.)

图三 昆明湖中古西堤桩基遗存（北京市颐和园管理处提供）

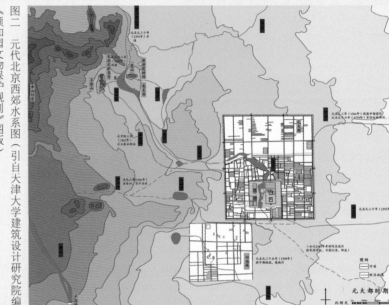

图二 元代北京西郊水系图（引自天津大学建筑设计研究院编《颐和园文物保护规划》图版）

图四 京杭道里图中的瓮山西湖（浙江省博物馆藏）

图五 清漪园建园前瓮山一带格局示意图（张龙. 颐和园样式雷建筑图档综合研究[D]. 天津：天津大学，2009.）

Garden in *Yuquanshan* (Jade Spring Mountain), and *Yuanmingyuan* were all constructed consecutively. Meanwhile, the original dyke of *Weng* Hill Lake was further reinforced to defend the low-lying *Changchun* Garden against flooding by the West Lake. The original dyke was then called the West Dyke because it was located on the west of *Changchun* Garden. According to literary records:

> The field on the east dyke of the West Lake covers thousands of hectares. *Yuanjing* Temple is built on the rocks at the foot of *Wengshan* hill. On the left side of the temple are green fields below and to the right side is a lake. This is where the most beautiful scenery around *Weng* hill begins. The West Lake is like a half moon when watched from Watching Lake Pavilion which is above the temple .

By combining archaeological findings (Fig.3) and in the description of *Weng* Hill in the Drawing of Beijing-Hangzhou Transport from the Kangxi Dynasty (Fig.4), the site before the construction of *Qingyiyuan* can be roughly understood (Fig.5).

2. The regulation of Mountains and Lakes

At the beginning of the Qing Dynasty, the construction of gardens to the west of Beijing meant that the demand for water was rising. Thus in Beijing itself access to water supplies became increasingly difficult — "The water in the moat was not even one *chi* deep (12.058 inches or 30.62 cm)". In order to solve the water deficits in the capital and to the west of the capital, The Qianlong Emperor personally carried out investigations and found that "Biyun Temple on the West Hill and the temples of *Xiangshan* all have dozens of famous springs with copious water. However, these springs mainly flow within these temples and monasteries and they disappear after flowing out of the hill area. Thus the area surrounding *Yuquan* Hill is flat and open, so ground water wells up from underground and forms a lake." The extension and desiltation of West Lake were begun in the 14th year of the reign of Qianlong (1749). At the same time, the waters from West Hills which emerge as the West Lake, were extended and then renamed, *Kunming* Lake the next year.

The extension of West Lake can be divided into two periods. One was the extension of the lake and desiltation to expand the lake's capacity during the winter of the 14th year of the reign of Qianlong (1749). The other period covers the extension of the West Lake to design gardens in the 15th year of Qianlong's reign (1750). The good orientation

---

家园林。在这些园林建设的同时，为使地势低洼的畅春园不受西湖泛滥的威胁，瓮山泊原有堤岸得

到进一步加固，因其位于畅春园迤西，被称为『西堤』。根据文献记载：

『西湖堤东稻畦千顷，接瓮山之麓，有寺曰圆静，因岩而构。』④

『瓮山圆静寺，左俯绿畴，右临碧浸，近山之胜于是乎始。』⑤

『其上为望湖亭，望西湖如半月。』⑥

结合考古发现（图三），以及康熙朝《京杭道里图》中对瓮山一带的描绘（图四），可大致勾

勒出清漪园建园之前的格局（图五）。

二、湖山整治

清初，随着北京西郊园林建设用水的不断增加，京城经常出现『城河水不盈尺』⑦的窘境。为

解决京师水荒与西郊水患，乾隆皇帝亲赴西郊考察，发现：『西山碧云、香山诸寺皆有名泉，其源

甚壮，以数十计，然惟曲注于招提蓝之内，一出山则伏流而不见矣，玉泉地就夷旷，乃腾迸而出，

潴为一湖。』⑧遂于乾隆十四年（1749年）冬启动西湖的拓展，清淤工程，同时铺设石槽将西山泉

水汇于拓展后的西湖，并于转年更名为『昆明湖』。

西湖的拓展可分为两个阶段，一是乾隆十四年（1749年）冬以扩容为主要目的拓湖、清淤；

二是乾隆十五年（1750年）以造园为主要目的的拓湖。瓮山、西湖良好的朝向，开阔的山水关系

及其在西郊诸园中的纽带位置（图六），都深深地打动了这位『山水之乐，不能忘怀』的乾隆皇帝，

图六 三山五园图（国家图书馆藏）

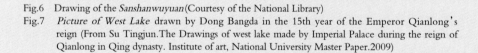

图七 乾隆十五年董邦达绘制的西湖图（引自苏庭筠.乾隆宫廷制作之西湖图.中央大学艺术学研究所硕士论文.2009）

Fig.6　Drawing of the *Sanshanwuyuan*(Courtesy of the National Library)

Fig.7　*Picture of West Lake* drawn by Dong Bangda in the 15th year of the Emperor Qianlong's reign (From Su Tingjun.The Drawings of west lake made by Imperial Palace during the reign of Qianlong in Qing dynasty. Institute of art, National University Master Paper.2009)

of _Weng_ Hill and West Lake, the open relationship between hills and rivers, and their reciprocal relations with different gardens in the west (Fig.6), all made a deep impression on the Qianlong Emperor who thought "The enjoyment provided by the Mountains and Rivers must not be forgotten." So he took the risk of being criticized and went back on his own words. He reconstructed _Qingyiyuan_ according to the principle of providing work opportunities rather than relief funds as a means of distributing money to people. Thus, the Qianlong Emperor set the scene for a succession of garden constructions during the first period.

First, he extended West Lake northwards to the foot of _Weng_ Hill, and eastward to _Haoshan_ Garden. These extensions changed the original situation in which there was no reciprocal relationship between the mountains and lakes. The mud from both the extension and desiltation of the lakes was used to increase the size of _Weng_ Hill, to alter its overall shape and to improve the soil conditions for growing trees.

Secondly, Qianlong kept the Dragon King Temple in the original dyke by making it into _Nanhu_ Island. This practice is not only a continuation of the historic context of the site, an enrichment of the garden scenery, and filling in of the space between the mountains and rivers, but also signalled the emperor's conception of constructing large scale gardens combining lakes, mountains, dykes and islands.

After the initial extension of _Kunming_ Lake, the Qianlong Emperor ordered the court painter, Dong Bangda, to paint _Picture of West Lake_ (Fig.7), and the emperor wrote a poem personally:

In the past, it was said that the West Lake was like a beauty. When her name was heard, her beauty would be known… I investigated Yan and Jin provinces in February and ordered painters to stay in Beijing and paint the Hangzhou landscape. When I came back, the drawing had been finished. It seemed that the landscape of Hangzhou was displayed on my desk. Ten scenes were laid out from the east to the west. Two mountains stood in the south and the north as if vying with each other for dominance… _Yuanmingyuan_ has rich sources of water, but no mountains. _Jingji Shanzhuang_ in Pan Mountain has great mountains, but no water. I should like to visit it (_Qingyiyuan_) because it is close by and I always long for its bright and beautiful lake scenery.

促使他甘冒自食其言⑨的非议，本着以工代赈，散财于民的原则再兴园工。因此，他在第一阶段就为后续园林建设埋下了伏笔。

首先，他将湖面向北拓展到瓮山脚下，向东拓展到好山园，改变了原来山湖偏离的关系，拓湖、清淤的泥土就近堆培瓮山，改造山形，改善绿化条件。

其次，保留原西堤上的龙王庙⑩，成就南湖岛。此举不仅延续了历史文脉，丰富了园林景观，填补了山水空间的空白，也透露乾隆皇帝要在此打造湖、山、堤、岛相结合的大型山水园的意象。

在昆明湖完成初步拓展后，乾隆皇帝命宫廷画师董邦达绘制《西湖图》（图七），并亲笔题诗：

『昔传西湖比西子，但闻其名知其美……岁惟二月巡燕晋，留京结撰亲承旨。归来长卷已构成，俨置余杭在案几。十景东西斗奇列，两峰南北争雄峙……淀池（圆明园）水富惜无山，田盘（盘山静寄山庄）山好拙于水。喜其便近每命游，具美明湖辄返企。』⑪

诗文中不仅点出了圆明园、静寄山庄山水不能兼备的遗憾，还透露了乾隆皇帝要在近郊写仿杭州西湖的想法。第二阶段的拓湖工程也随之展开。

与杭州西湖相比，第一阶段工程完成后，还缺少两水之间的长堤。于是，乾隆皇帝在原有西堤西北段的基础上，根据景观需要向南延展，成为昆明湖的西堤，同时在其西侧挖湖、置岛。正如其在御制诗文中所述：

『面水背山地，明湖仿浙西。琳琅三竺宇，花柳六桥堤。』⑫

The poem not only highlights the emperor's regret that neither *Yuanmingyuan* nor *Jingji Shanzhuang* has both mountains and water, but also reflects Qianlong's wish to make a copy of the West Lake of Hangzhou in the nearby suburbs. The second period of extension of the West Lake was launched after Qianlong wrote the poem.

Compared with the West Lake of Hangzhou, after the first period of constructing the second lake, there was no long dyke between the two lakes. Thus, to improve the scenery and make it like Hangzhou, the Qianlong Emperor extended the dyke on the northwest of the original lake southwards so that it became the west dyke of *Kunming* Lake. Meanwhile, he deepened the lake and formed the island on the west side. As described in his poem:

This place faces the lake and backs on to the mountain. The *Kunming* Lake imitates the West Lake in Zhejiang. The Buddhist temples are like beautiful jade. The willows and flowers decorate the dykes of the bridges.

Qianlong also constructed Little *Xiling* Island on the west side of Longevity Hill by copying the relationship between *Gu* Hill and North Hill around the West Lake in Hangzhou. He kept *Zhichun* Pavilion Island to blend with the scenery from *Yuquan* Hill and *Yufeng* Pagoda. He extended *Weng* Hill lake around the western end of the former *Weng* Hill to the back making a series of rivers and lakes so as to form a backdrop reflecting Longevity Hill from all sides. Through a succession of major landscape changes, scenery was formed of "Open hills encircled by elegant rivers; scenes more beautiful than *Penglai* [Realm of the immortals]" (Fig. 8, Fig.9).

3. The Construction of Architecture

After the formation of hills and lakes, how can there be no pavilions and terraces? To celebrate the 60th birthday of his mother, Empress Dowager Chongqing, the Qianlong Emperor constructed the Buddhist temple complex called the Great Temple of Gratitude and Longevity (*Da bao'en yanshou si*) on the original site of *Yuanjing* Temple in *Weng* Hill. "Weng Hill" was renamed "Longevity Hill". Through these activities, the hill was transformed into an ideal Buddhist site, conveying the emperor's wish to pray for his mother to live longer and having good fortune. These temples included: *Cifu* Pavilion, *Luohan* Hall, *Baoyun* Pavilion, *Zhuanlunzang*, *Xumilingjing*, *Yunhui* Temple, *Shanxian* Temple, *Wusheng* Temple, *Huacheng* Pavilion, *Tanhua* Pavilion, and others. Other architecture serving to aesthetically enhance the beautiful landscape of lakes and hills

复回，风流文采胜蓬莱』的山水胜景。（图八、图九）

伸，形成了后河后湖，与万寿山呈迂回合抱之势。经过不断的深化改造，逐步构成了『秀水明山抱

景玉泉山玉峰塔特意预留知春亭小岛，为了加强湖山的整体联系，又将湖水沿瓮山西麓向山后延

另外，乾隆皇帝又通过对西湖孤山与北山关系的写仿，在万寿山西侧堆出小西泠岛[十三]；为借

三、建筑营造

山水既成，岂能无亭台之点缀[十四]。为庆祝其生母崇庆太后的六十万寿，乾隆皇帝在瓮山圆静寺

旧址兴建大报恩延寿寺，并将『瓮山』定名为『万寿山』，旨在通过庙宇的创建，营造浓厚的宗教

山林气氛，为母亲祈福延寿。这些庙宇还有慈福楼、罗汉堂、宝云阁、转轮藏、须弥灵境、云会寺、

善现寺、五圣祠、花承阁、昙花阁等。其他功能性以及点缀湖山的建筑，也在乾隆皇帝具体而微的

设计要求与指导下陆续展开。如乾隆十六年（1751 年），精简后的织染机房和相关职员的住房怎么营建？怎么与园林创

已有的水田景观共同构成『耕织图』景观，但织染机房和相关职员的住房怎么营建？怎么与园林创

作相结合？当年的一份相关《奏销档》记录了乾隆皇帝的具体批示：

『织染局移到万寿山附近……实难早晚应候官差，仰懋圣恩，每人各赏给官房一间……共盖造

小房八十余间，每人赏房一间。但此项房间若盖连房，似觉未宜，请交该工于局作附近地方，合其

形势，或二三间、三四间不等，布成村落，以标幽致，即于该匠役房间空间之地，种植桑林以养丝

蚕，如此则匠役等既得楼止之地，而村居蚕桑点缀于山水之间，益着园亭之盛也。』[十五]

图九　颐和园全景（北京市颐和园管理处提供）

图八　西湖全景（周兔英摄．引自：西湖申遗文本）

Fig.8　The whole scenery of West Lake in Hangzhou (Photo by Zhou Tuying, from: The Documents for the Application for West Lake to be Included in the World Heritage List)

Fig.9　The scenery of the Summer Palace (Supplied by the Administrative office of the Summer Palace, Beijing)

was also initiated according to the detailed requirements and orders of the Emperor Qianlong . For example, in the 16th year of Qianlong's reign (1751), when the number of workers in the Weaving Workshop was reduced , the workshop and the workers were moved to the west side of Longevity Hill, where both workers and workshop were integrated with the paddy field scenery to form the Farming and Weaving Picture Scenic Area. But how were the weaving workshop and weavers' homes constructed ? How were they integrated with the creation of the garden landscape? Qianlong's solutions to these questions were recorded in detail in the "Accounting reports of the Inner Court" (*Neiwufu zouxiao dang*):

The Weaving Bureau was moved to Longevity Hill…It was difficult for the workers to serve the officials on a daily base. Through the favour of the emperor, each of the workers was awarded a room here…There were over 80 small houses. Everyone got one. Had such houses been reconstructed in the shape of linear housing blocks, comprising sets of two to three houses or three to four houses all together, they would appear as villages. Such an appearance did not become a scenic area. Rather the sets of houses should be constructed separately each surrounded by mulberry trees planted both to feed the silk worms and to form arboreal 'green-space' between houses. Such a landscape design provided the officials with dwelling space and the mulberry clad villages could greatly enhance the mountain landscape.

Architecture of the period in *Qingyiyuan* may be categorized according to their practical functions, cultural and landscape conceptions. Some buildings function as the gates of entrances to gardens, such as East Palace Gate, West Palace Gate, and North Palace Gate. The Hall of Benevolence and Longevity functioned as a temporary court building. Oriole-Listening Hall was used for entertainment; Buddhist temples embodied the culture of Buddhism; Daoist temples conveyed Taoist culture, such as *Wusheng* Temple, Dragon King Temple, God of Flower Temple, etc; Scholar Gardens symbolized Confucian culture, such as *Huishan* Garden, *Gaichun* Garden, *Changguan* Hall, *Qiwang* Pavilion etc; Entrance buildings to the city symbolize the city, such as *Wenchang* Gate Pavilion, *Suyunyan* Pavilion, *Yinhui* Gate Pavilion etc; Water Village symbolized the countryside; Shopping Street symbolized commerce; Weaving Workshop symbolized industry, etc. (Fig.10). These varied types of symbolic architecture reflect design concepts "*Zhoubi* (heaven) and *Ying* sea are so vast, Mountain *Kunlun* and Island *Fanghu* are accommodated here [in my garden]", "Don't distinguish things of the East or West, just use them to decorate the noble garden" of Qianlong.

清漪园时期的这些建筑如果按功能、文化和景观意象来划分，则有作为园林出入口的宫门，如东宫门、西宫门、北宫门；承担临时办公的宫殿——仁寿殿；满足观演功能的听鹂馆戏楼；彰显佛教文化的寺庙，如大报恩延寿寺、须弥灵境等；体现道教文化的祠庙，如武圣祠、龙王庙、花神庙等；象征儒家文化的文人园，如惠山园、贻春园、畅观堂、绮望轩等；呈现城市意象的城关，如文昌阁、宿云檐、寅辉城关等；表现乡村意象的水村居，显示商业意象的买卖街，代表手工业意象的织染局（图十）……建筑类型与意象之丰富，正是乾隆皇帝『周裨瀛海诚旷哉，昆仑方壶缩地来』[4]、『何分西土东天，倩他装点名园』[5]等造园思想与理念的直接体现。

图十 乾隆朝清漪园建筑格局示意图（张龙．济运疏名泉·延寿创刹宇——乾隆时期清漪园山水格局分析及建筑布局初探[D]．天津：天津大学，2006．）

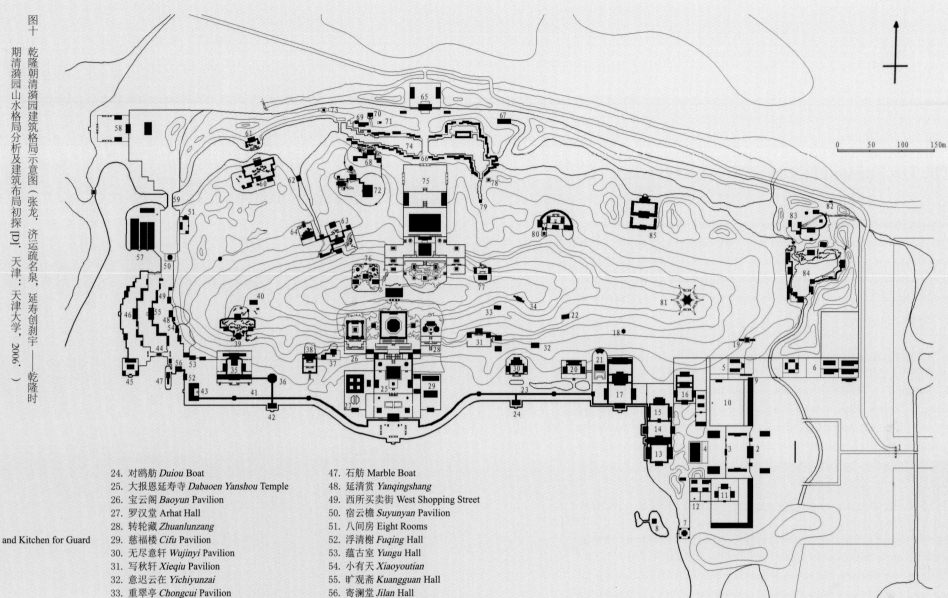

1. 涵虚牌楼 Hanxu Pailou
2. 东宫门 East Palace Gate
3. 二宫门 Second Palace Gate
4. 勤政殿 Qinzheng Hall
5. 茶膳房 Kitchen
6. 外膳房、侍卫饭房 Outside Kitchen and Kitchen for Guard
7. 文昌阁 Wenchang Gate Pavilion
8. 知春亭 Zhichun Pavilion
9. 进膳门 Gate for Delivering food
10. 进膳区 Area for Dining
11. 军机处 Office of the Grand Council
12. 耶律楚材祠 Ancestral Temple of Yelv Chucai
13. 玉澜堂 Yulan Hall
14. 夕佳楼 Xijia Pavilion
15. 宜芸馆 Yiyun Hall
16. 怡春堂 Yichun Hall
17. 乐寿堂 Leshou Hall
18. 含新亭 Hanxin Pavilion
19. 赤城霞起 Chichengxiaqi Pavilion
20. 养云轩 Yangyun Hall
21. 乐安和 Leanhe
22. 餐秀亭 Canxiu Pavilion
23. 长廊东段 East Section of the Long Corridor

24. 对鸥舫 Duiou Boat
25. 大报恩延寿寺 Dabaoen Yanshou Temple
26. 宝云阁 Baoyun Pavilion
27. 罗汉堂 Arhat Hall
28. 转轮藏 Zhuanlunzang
29. 慈福楼 Cifu Pavilion
30. 无尽意轩 Wujinyi Pavilion
31. 写秋轩 Xieqiu Pavilion
32. 意迟云在 Yichiyunzai
33. 重翠亭 Chongcui Pavilion
34. 千峰彩翠 Qianfengcaicui Gate Pavilion
35. 听鹂馆 Tingli Pavilion
36. 山色湖光共一楼 Shansehuguang-gongyilou
37. 云松巢 Yunsongchao
38. 邵窝 Shaowo
39. 画中游 Huazhongyou
40. 湖山真意 Hushanzhenyi
41. 长廊西段 West Section of the Long Corridor
42. 鱼藻轩 Yuzao Pavilion
43. 石丈亭 Shizhang Hall
44. 荇桥 Xing Bridge
45. 五圣祠 Wusheng Temple
46. 水周堂 Shuizhou Hall

47. 石舫 Marble Boat
48. 延清赏 Yanqingshang
49. 西所买卖街 West Shopping Street
50. 宿云檐 Suyunyan Pavilion
51. 八间房 Eight Rooms
52. 浮清榭 Fuqing Hall
53. 蕴古室 Yungu Hall
54. 小有天 Xiaoyoutian
55. 旷观斋 Kuangguan Hall
56. 寄澜堂 Jilan Hall
57. 北船坞 North Boatyard
58. 如意门（西宫门）Ruyi Gate(West Palace Gate)
59. 半壁桥 Banbi Bridge
60. 绮望轩 Qiwang Pavilion
61. 看云起时 Kanyunqishi
62. 澄碧亭 Chengbi Pavilion
63. 赅春园 Gaichun Garden
64. 味闲斋 Weixian Hall
65. 北楼门 North Palace Gate
66. 三孔石桥 Three-arch Bridge
67. 后溪河船坞 Back River Boatyard
68. 绘芳堂 Huifang Pavilion
69. 嘉荫轩 Jiayin Pavilion

70. 妙觉寺 Miaojue Temple
71. 花神庙 Huashen Temple
72. 构虚轩 Gouxu Pavilion
73. 通云 Tongyun Pavilion
74. 后溪河买卖街
    Back River Shopping Street
75. 须弥灵境 Xumilingjing
76. 云会寺 Yunhui Temple
77. 善现寺 Shanxian Temple

78. 寅辉城关 Yinhui Gate Pavilion
79. 南方亭 South Square Pavilion
80. 花承阁 Huacheng Pavilion
81. 昙花阁 Tanhua Pavilion
82. 东北门 Northeast Gate
83. 霁清轩 Jiqing Pavilion
84. 惠山园 Huishan Garden
85. 云绘轩 Yunhui Pavilion

Fig.10　Diagram of the architectural structure of Qingyiyuan in Qianlong
(ZHANG Long. Analysis of Landscape Pattern and Research of Architectural Layout of Qingyi Garden during Qianlong Reign[D]. Tianjin: Tianjin University,2006.)

The Decline and Destruction of *Qingyiyuan*

In the final years of Qianlong's reign, the Qing Dynasty reached its peak and started to decline. Both internal and external problems followed. Successive emperors had neither time nor energy to take care of the construction of gardens. Thus, the existing gardens lost their past glories: some were closed, others were distributed among princesses and princes, furniture was removed, and buildings that could not be repaired were demolished. *Qingyiyuan* was no exception.

1. Adjustments during the Jiaqing period (1796—1820)

According to the record of known documents, except for the regular repair of *Qingyiyuan* during the Jiaqing period, there were only two large scale alterations. One was that *Huishan* Garden was renamed the Garden of Harmonious Pleasures and parts of the buildings were added or adjusted. The other was that the three-story *Wangchan* Pavilion on *Nanhu* Island was changed into the single-storey *Hanxu* Hall.

2. Decline during the Daoguang Period (1821—1850)

After the accession of the Daoguang Emperor, he strongly promoted austerity. Although he often went to *Qingyiyuan*, he mostly prayed for rain in *Guangrun Lingyu* Temple. The record that often appeared in the relevant documents is that the furnishings in the architectural buildings in *Qingyiyuan* were removed, caught fire, or collapsed. For instance, because of cuts in administrative costs and the destruction of buildings, the Daoguang Emperor removed the furnishings from *Shuizhou* Hall, *Le'an'he*, *Zaojian* Hall, *Yunhui* Pavilion, *Gouxu* Pavilion, *Qinzheng* Hall, *Kuangguan* Studio, and Stone Boat, etc.. *Gouxu* Pavilion and *Yichun* Hall caught fire in the 18th and 24th year of the Emperor Daoguang's reign. The Temple of Fragrant Solemnity (*Xiangyan zongyin Pavilion*) collapsed because the major wooden structure rotted in the 30th year of Daoguang's reign (1850).

3. Disaster during the Xianfeng Period (1851—1861)

In the tenth year of the Xianfeng Emperor's reign (1860), the British and French destroyed the imperial gardens in the west of Beijing, including *Qingyiyuan*. According

清漪园衰落与焚毁

乾隆末年，清王朝盛极而衰，内忧外患接踵而至，嗣后继位的帝王也无暇于新的园林建设活动，已有的园林也逐渐失去了往日的光辉，或封存，或分封其他王爷、公主，或撤去陈设，或拆除无力维修的建筑，清漪园也不例外。

一、嘉庆朝（1796—1820 年）的微调

根据已知档案记载，嘉庆朝清漪园除常规维修外，主要有两次规模相对较大的改动，一是惠山园更名为『谐趣园』，并添改部分建筑；二是将南湖岛上三层的望蟾阁改为单层的涵虚堂。

二、道光朝（1821—1850 年）的衰落

道光即位后，力崇节俭，虽常至清漪园，也多是到广润灵雨祠拈香求雨，在相关档案中更多出现的则是清漪园建筑裁撤陈设、不幸于火或坍塌的记录。或为节省管理开支，或因建筑残毁，道光皇帝先后撤去了水周堂[18]、乐安和[19]、藻鉴堂[20]、云绘轩[21]、构虚轩[22]、勤政殿[23]、旷观斋[24]、石舫[25]等处陈设。构虚轩、怡春堂分别于道光十八年（1738 年）[26]、二十四年（1744 年）[27]被烧毁。道光三十年（1850 年）香岩宗印之阁因大木糟朽而坍塌[28]。

to comprehensive analysis of the documents and photos after the destruction (Fig.11), most of the wooden buildings on Longevity Hill and on the east bank of *Kunming* Lake were ruined. Only some small courtyards, side halls, and wooden buildings on the west of *Kunming* Lake and bridges, pagodas, and pailou (Chinese archways) with their stone or glass structures, survived the disaster.

4. The Abandonment of the Tongguang Period (1862—1886)

During the reign of the Tongzhi Emperor and the Guangxu Emperor, officials were still selected for *Qingyiyuan* according to convention and they were also ordered to burn incense regularly in the Dragon King Temple. In the eighth month of the 12th year of his reign, the Tongzhi Emperor planned to reconstruct *Yuanmingyuan* to praise the Empress Dowagers with filial piety. Tongzhi had the rotten parts of the timber stored in *Qingyiyuan*, *Jingyi* Garden, and *Jingming* Garden picked out and then had them processed into smaller sizes for use in construction. At the same time, he investigated the decayed buildings of the three gardens. The wood however, was so rotten and small that it couldn't be used for reconstruction. From the 1860 disaster during Xianfeng's reign to the reconstruction in the 12th year of the reign of the Guangxu Emperor (1886), *Qingyiyuan* naturally decayed and most of the buildings collapsed during theoe 25 years.

三、咸丰朝（1751—1861年）的浩劫

咸丰十年（1860年）英法联军焚掠北京西郊皇家园林，清漪园同时罹难，综合分析相关档案和焚毁后的老照片（图十一），可知万寿山及昆明湖东岸体量较大的木构建筑多数被毁，仅有部分体量较小的院落、配殿和远在昆明湖西部的木构建筑，以及砖石、琉璃结构的石桥、佛塔、牌楼等幸免于难。

四、同治、光绪朝（1862—1886年）的荒废

同治、光绪朝清漪园仍照例选派官员，定期遣官诣龙神庙拈香。同治十三年（1874年）八月，亲政后的同治皇帝为孝养两宫皇太后，计划重修圆明园，将清漪园、静宜园、静明园三处所存糟朽木植，剔出糟朽，改成小件使用。同时勘查三园坍塌、歪闪房屋，虽因其木植糟朽，尺寸较小，不堪选用而放弃拆用[二十九]，但从咸丰朝罹难，至光绪十二年（1886年）前重修，历时二十五载，清漪园经风雨摧残，已大数衰败。

图十一　焚毁后的大报恩延寿寺（北京市颐和园管理处提供）

Fig.11　The DabaoenYanshou Temple which was burned down (Supplied by the Administrative office of the Summer Palace, Beijing)

Reconstruction of Summer Palace

In the 12th year of the reign of the Guangxu Emperor when he was 16 years old, he took over the reins of leadership. The Empress Dowager Cixi planned to resign from her leadership position and construct her own palace. After the Tongzhi Emperor's failure in reconstructing the *Yuanmingyuan*, Cixi this time chose *Qingyiyuan*. This was not only because that the scenic structure of *Qingyiyuan* was the most resplendent (Fig.12), but also because she aimed to restore the original function of the garden as a node for the water transport system for Beijing. In order to avoid causing a shortage of funds with the restoration project being launched at once, which would provoke criticism both at court and with the public, Cixi, with the full support of Yixuan (Lord Chunqin), Minister of Navy *Yamen* and father of the Guangxu Emperor, launched the restoration project and implemented the project in stages all under the rubric of developing the 'Eight Banners' navy.

1. Construction of the Water Exercise School

Through a series of failures in foreign wars, the Qing Government gradually realized the importance of establishing new style of navy. In the ninth month of the eleventh year of Guangxu's reign (1885), the Navy *Yamen* was set up and Lord Chunqin governed all naval affairs. On the 27th day of the eight month of the following year (1886), Yixuan requested to resume the *Kunming* Water Exercise. At the same time, he also announced that in order to allow the Dowager Cixi to watch the naval exercises, he planned to repair the original palaces, pavilions and the halls on Longevity Hill and halls in the Temple of *GuangruLingyu* and the bridges, *Pailou* along the lake. This request not only accorded with the needs of the situation, but was also was good for launching the reconstruction. Thus, Cixi immediately approved the request and gave directions to "recover the external and internal water exercise halls of *Kunming* Lake according to Western methods" in order to train the 'Eight Banners' navy.

From the day of Yixuan's request, the Navy *Yamen* constructed the external and internal water exercise halls separately in the Water Village of *Qingyiyuan* and in the Farming and Weaving Picture Scenic Area (Fig.13). The construction was finished before the opening of the school on the 15th day of the twelfth month of the 13th year of Guangxu's reign (1887).

颐和园的重修

光绪十二年（1886年），皇帝年满十六岁，到了『亲政』的年龄，慈禧太后计划退居二线，营建自己归政后的御苑。经历了同治朝重修圆明园工程的失败，慈禧太后此次选择了景观格局最具皇家气派（图十二），同时承担京城水利枢纽的清漪园作为重修目标。为避免重修项目集中实施导致经费紧张，进而引发朝野非议，在海军衙门总理大臣、光绪皇帝生父醇亲王奕譞的倾力支持下，重修工程以恢复『昆明水操』，以培养八旗海军人才的名义启动，并有计划地分批实施。

一、水操学堂建设

经历晚清一系列对外战争的失利，清廷逐渐意识到建立新式海军的重要性，光绪十一年（1885年）九月成立海军衙门，以醇亲王奕譞总理海军事务。光绪十二年（1886年）八月二十七日，奕譞奏请恢复昆明水操，同时声明为恭备皇太后阅操，拟将万寿山暨广润灵雨祠旧有殿宇台榭，并沿湖各桥座、牌楼酌加保护修补以供临幸。⊙此请既符合时局需要，又利于园工开展，慈禧当即准其奏请，并指示『参用西法，复昆明湖水操内外学堂』⊙⊙，用以专门培训八旗子弟。

自奏请之日起，海军衙门就分别在清漪园水村居和耕织图的基址上修建水操内外学堂（图十三），工程于光绪十三年（1887年）十二月十五日开学前竣工。

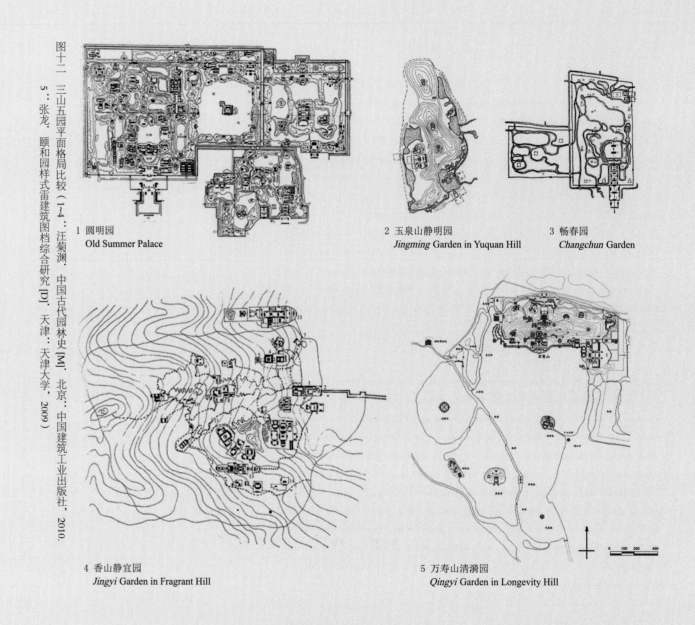

图十二　三山五园平面格局比较（1~4：汪菊渊．中国古代园林史 [M]．北京：中国建筑工业出版社，2010.
5：张龙．颐和园样式雷建筑图档综合研究 [D]．天津：天津大学，2009）

1　圆明园
Old Summer Palace

2　玉泉山静明园
Jingming Garden in Yuquan Hill

3　畅春园
Changchun Garden

4　香山静宜园
Jingyi Garden in Fragrant Hill

5　万寿山清漪园
Qingyi Garden in Longevity Hill

Fig.12　Comparison of the master plans of the Three Mountains and Five Gardens (1~4:WANG Jvyuan. *History of Ancient Chinese Gardens*[M].Beijing: China Architecture and Building Press,2010.
5:ZHANG Long. *Comprehensive Research on Yangshi Lei Archives of Summer Palace*[D]. Tianjia: Tianjin University, 2009)

图十三　水操学堂格局图（国家图书馆藏）

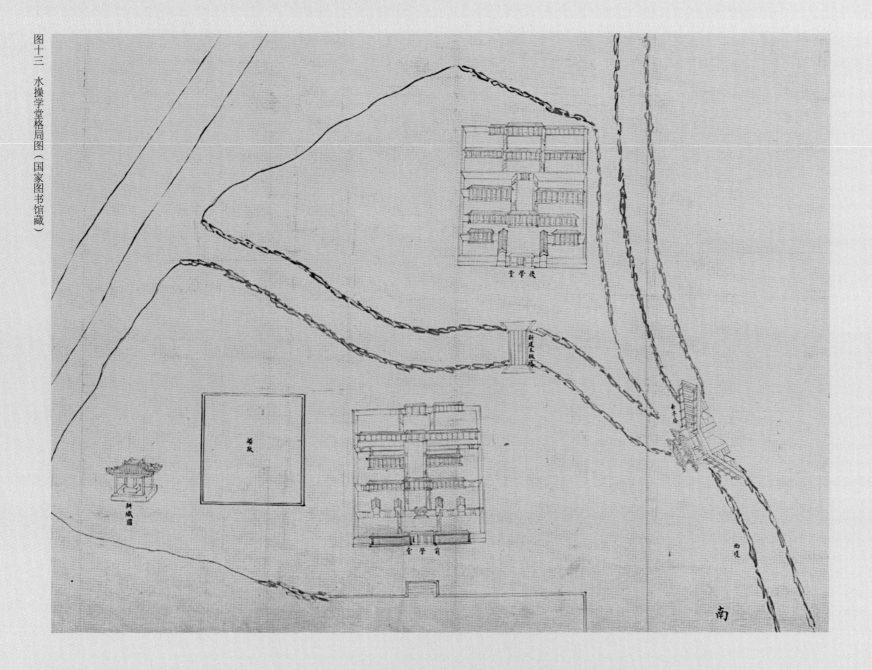

Fig.13　The Layout of the Water Exercise School (Collection of the National library)

2. Construction of *zhenglu* (the temple halls along the central path) and *foyu* (temple palace)

According to the records such as *Opening of the School of Water Exercise and Conditions for Supplying Beams for the Hall of Dispersing Clouds* — on the same day as the opening of the School of Water Exercise, ceremonies for supplying beams were held separately at the *Paiyun* Hall (Hall of Dispersing Clouds), *Dehui* Hall (Hall of Virtuous Light), and at the temples on the back of the hill (*Xiangyan Zongyin* Pavilion, or Temple of Fragrant Solemnity). In the tenth month of the 14th year of the reign of Emperor Guangxu, the *Zhendu* Gate of the Forbidden City caught fire. This accident provided an excuse for the ministers and workers who were against the restoration project. Under pressure from public opinion, Empress Dowager Cixi issued an edict on the 26th day of the twelfth month in the same year: "All construction on the Summer Palace will cease as an austerity measure, except for the temple palace (*foyu*) and the temple halls along the central path (*zhenglu*). *Foyu* and *zhenglu* refer to the Hall of Dispersing Clouds, Hall of Virtuous Light on the front hill, and temple halls on the back hill (*Xiangyan Zongyin* Pavilion). Because of the importance of the Hall of Dispersing Clouds, Lei Tingchang designed many schemes and did detailed surveys (Fig.14, Fig.15) According to the records of the *Construction List* of the Summer Palace, these constructions lasted until 1895, the 21st year of Guangxu's reign.

3. Project of Watching Exercises

To avoid provoking ministers and workers who disapproved of the entire reconstruction of the Summer Palace, Cixi did not allow Lei Tingchang to launch the entire design of reconstruction, but rather gave priority to "the project for Dowager Cixi to watch the naval exercises" that is referred to in the agenda for the meeting at court as "resumption of water exercises." In the 13th year of the reign of Guangxu (1887), the *Construction List of the repaired and unrepaired architectures on Longevity Hill* lists most constructions, including the East Gate of the Palace, the Gate Area of *Renshou* Palace and *Wenchang* Pavilion, *Yulan* Hall, *Leshou* Hall, Long Corridor, *Nanhu* Island, *Kuoru* Pavilion around the lake. This part of the construction was basically finished before the emperor and empress visited the garden on the 23rd of the third month of the 15th year of Guangxu's reign (1889).

4. Project of 'Taking good care of health'

In order to legitimize the restoration project, the Guangxu Emperor announced on the first

---

二、『正路』『佛宇』工程

据中国第一历史档案馆所藏《水操学堂开学、排云殿供梁情况折》等档案记载，就在水操学堂开学的同一天，排云殿、德辉殿、后山佛殿分别举行了供梁仪式[二十一]。光绪十四年（1888年）十月，紫禁城贞度门失火，这一事件给反对重修颐和园的臣工提供了口实，迫于舆论压力，慈禧太后于是年十二月二十六日颁布懿旨：『所有颐和园工程，除佛宇及正路殿座外，其余工作一律停止，以昭节俭』[二十二]。其中所说的『佛宇』与『正路』工程即前山排云殿、德辉殿和后山佛殿（香岩宗印之阁）。鉴于排云殿建筑群的重要性，雷廷昌设计了多个方案，进行反复推敲（图十四、图十五）。据颐和园《工程清单》[二十四]记载，这些『佛宇、正路』工程一直持续到光绪二十一年（1895年）。

三、阅操工程

为避免因全面展开设计工作而引起臣工反对重修颐和园，慈禧太后并未让雷廷昌全面启动设计工作，而是优先设计『阅操工程』，即『恢复水操』奏折上提及的那些恭备皇太后阅操而修的工程。光绪十三年（1887年）《万寿山等处已修齐未修齐工程清单》[二十五]大致开列了这些工程，主要包括东宫门、仁寿殿门区、文昌阁、玉澜堂、乐寿堂、长廊、南湖岛、廓如亭等环湖建筑。这部分工程于光绪十五年（1889年）三月二十三日慈禧太后和光绪帝来园阅操前基本完成。[二十六]

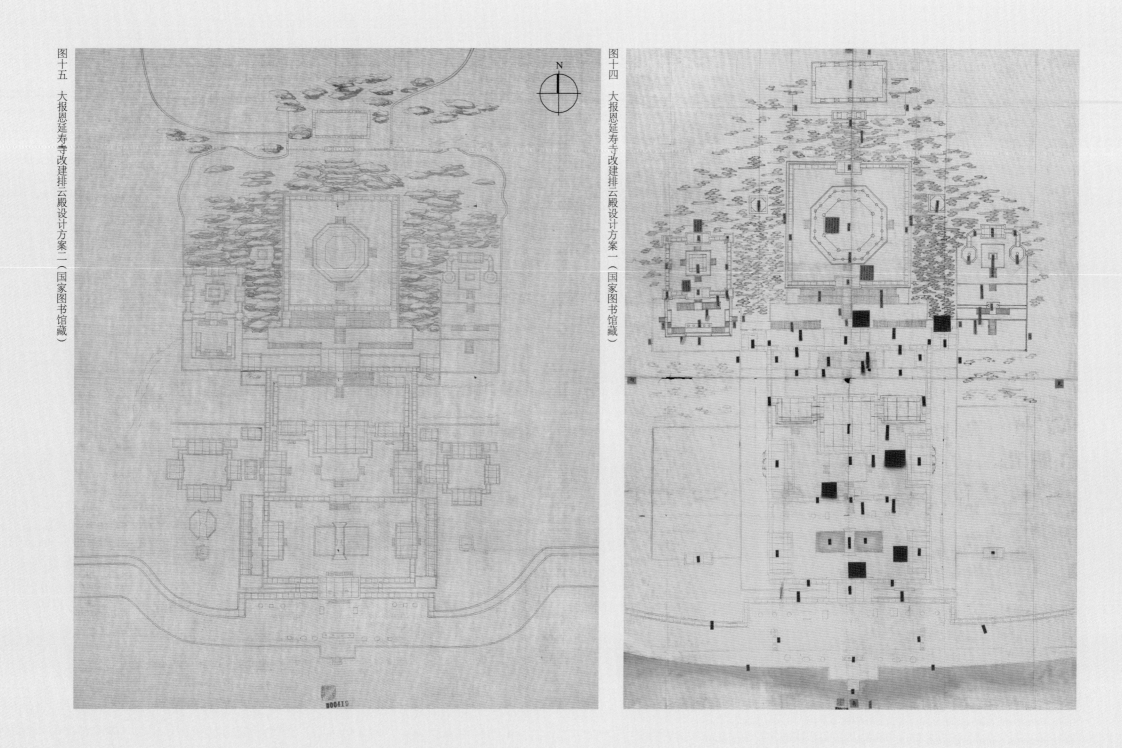

图十五　大报恩延寿寺改建排云殿设计方案二（国家图书馆藏）

图十四　大报恩延寿寺改建排云殿设计方案一（国家图书馆藏）

Fig.14　Schemes 1 for transforming the Great Temple of Gratitude and Longevity into the Hall of Dispersing Clouds (Collection of the National library)

Fig.15　Schemes 2 for transforming the Great Temple of Gratitude and Longevity into the Hall of Dispersing Clouds (Collection of the National library)

day of the second month of 1888, the 14th year of Guangxu's reign:

The Great Temple of Gratitude and Longevity on Longevity Hill is the place where Emperor Qianlong celebrated his mother's birthday three times. It is particularly appropriate to follow fully previous practices and regulations with reverence. *Qingyiyuan* was renamed Summer Palace–Garden for taking care of health and harmonization. All the architectural buildings were repaired for the empress. In the year of great celebration, I shall personally lead all the officials in prayer for my mother so as to express my filial obligations and admiration.

The rationale that Emperor Guangxu followed the example set by the Qianlong Emperor caring for the Empress Dowager is appropriate. At the same time, Guangxu stressed that he was only repairing a small section, to get the approval of the public. As is known in the edict , this construction was for taking good care of Cixi's health after she retired, thus the project was entitled 'Taking good care of health'. This construction was divided into two types. One type was to build new buildings: to meet the emperor and empress's living requirements, to hold a court for governance, to provide entertainment, and other subsidiary buildings such as *Shengpingshu* (Performances Office), the imperial kitchen, *Shoushanfang* (kitchen for the birthday banquet), offices of different departments outside the east gate, and the rebuilt *Dehe* Garden in the original location of *Yichun* Hall (Fig.16). The other type was those constructions giving consideration to scenery, economy and practicality. For example, the *Huazhongyou* group of architectural buildings was an important example of scenic architecture on the west side of Longevity Hill. The principal building, the three-story *Chenghui* Pavilion, was the highest on the west side of Longevity Hill. The restoration scheme largely kept the original form of this group of buildings. The three-storey *Canxiu* Pavilion, located on the east side of the front of Longevity Hill, had great view. Despite the limited budget, they did not merely reduce the three storeys to just a one-storey building. Rather they built a one-storey building, *Fuyin* Pavilion (Fig.17), but designed it in the form of an open book with a flat roof used as a terrace for viewing. Both sides of the book-shaped building had rock shaped stairs to reach the roof. This move not only continued the functions of the *Canxiu* Pavilion's scenic look out, but also enriched the garden architecture as a result of its book-shaped plan form; The three-storey six pointed-star shaped *Tanhua* Pavilion on the ridge on the east side of Longevity Hill (Fig.18) was designed as two schemes following the original three storey design, but considering the austerity both were reduced to a one-storey building (Fig.19, Fig.20). Eventually, they chose a practical form for the *Jingfu* Pavilion, i.e. one storey with three

四、颐养工程

为了使重修工程合法化，光绪十四年（1888年）二月初一光绪皇帝发布上谕：

『万寿山大报恩延寿寺，为高宗纯皇帝（乾隆）侍奉孝圣宪皇后三次祝嘏之所，敬踵前规，尤征详恰。其清漪园旧名，谨拟改为颐和园。殿宇一切，亦量加葺治，以备慈舆临幸。恭逢大庆之年，朕躬率群臣，同申祝悃，稍尽区区尊养微忱。』

以孝养太后为由，效仿乾隆皇帝，可谓名正；同时强调，『量加修葺』，以获天下臣民认可。

由上谕可知这批工程是为慈禧归政后颐养所修，故称之为『颐养工程』。这批工程大致可分为两种类型，一是为满足帝后园居、理政、娱乐等需求的新建和改建，如东宫门外新建升平署、御膳房、寿膳房、各部公所等大量附属建筑，以及在怡春堂旧址改建的德和园（图十六）。二是为兼顾景观、经济和实用等要求对清漪园时期点景建筑进行复原和改建，尤其是三层的主体建筑澄辉阁，是西区的制高点，重修方案基本保留了原有的形式；位于万寿山前山东麓三层、可登临的餐秀亭，观景效果极佳，虽囿于经济因素，改建为一歇山卷棚敞厅了事，而是选择了平顶书卷形的福荫轩（图十七），其两侧有楼山可登顶，此举既延续了餐秀亭的观景功能，书卷形的平面格局也丰富了园林建筑类型；万寿山东麓山脊上三层六角星形的昙花阁（图十八），最初从经济角度考虑，设计了两个忠实原作平面，但改为一层的方案（图十九、图二十），最后在综合技术、经济和实用等因素后，选择了一个实用性强的单层三卷勾连搭的景福阁（图二十一）。

021

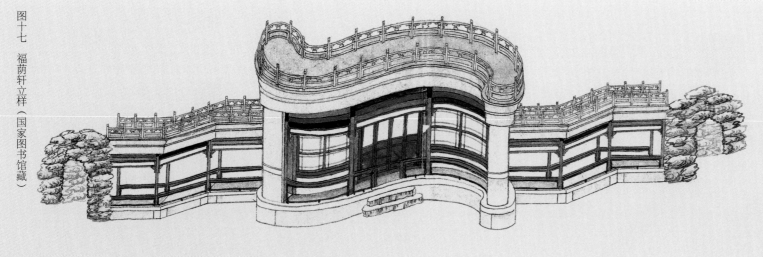

图十七　福荫轩立样（国家图书馆藏）

图十六　怡春堂改修德和园地盘样（国家图书馆藏）

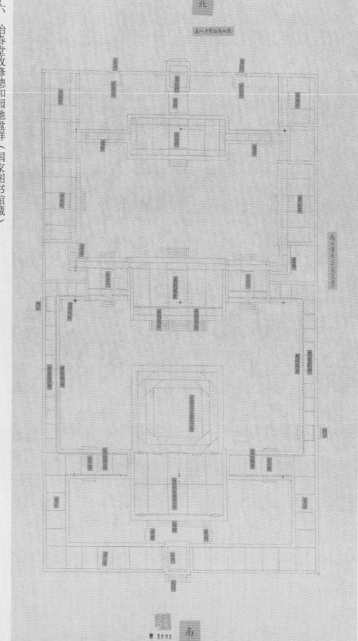

图十八　清漪园时期的昙花阁（1860年比托拍摄）（引自Harris, D.Of Battle and Beauty: Felice Beato's Photographs of China[M]. Santa Barbara,CA: Sannta Barbara Museum of Art, Berkeley: University of California Press, 1999）

Fig.16　Site Plan for transforming *Yichun* Hall into *Dehe* Garden (Collection of the National library)
Fig.17　Elevation of *Fuyin* Pavilion (Collection of the National library)
Fig.18　*Tanhua* Pavilion during the period before the reconstruction of *Qingyiyuan*, photoed by Beato in 1860 (From Harris, D.Of Battle and Beauty: Felice Beato's Photographs of China[M]. Santa Barbara, CA:Sannta Barbara Museum of Art, Berkeley: University of California Press, 1999)

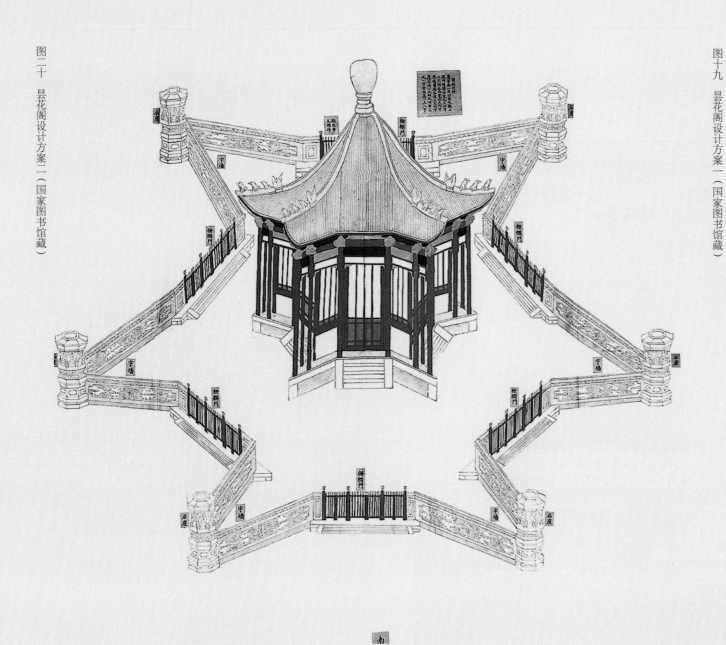

图二十　昙花阁设计方案二（国家图书馆藏）

图二十九　昙花阁设计方案一（国家图书馆藏）

北

南

北

南

Fig.19　Schemes 1 for the reconstruction of *Tanhua* Pavilion (Collection of the National library)

Fig.20　Schemes 2 for the reconstruction of *Tanhua* Pavilion (Collection of the National library)

图二十一　景福阁（作者自摄）

Fig.21　*Jingfu* Pavilion (Photographed by author)

connected roofs reconstructed giving consideration to technical and economic factors (Fig.21).

According to the *Basic Volume of Rules, Money and Food* and the *Detailed Volume of Materials and Money* put out by the designer's office and the accounting office, the project of 'Taking good care of health' included the Great Opera Hall of *Dehe* Garden, the Tower of Buddhist Fragrance, the Garden of Harmonious Pleasures and many subsidiary architectural buildings outside the east gate. It is recorded in the *Construction List* of the Summer Palace that "Construction for the project of 'Taking good care of health' lasted until 1895, the 21st year of Guangxu's reign.

5. Subsequent Construction

In 1892, the 18th year of Guangxu's reign, as the project of 'Taking good care of health' had not been completely finished, the Empress Dowager Cixi began to live for a short-term in the Summer Palace. The failure of China's naval engagement in the first Sino-Japanese War in 1895 (the 21st year of Guangxu's reign) did not shatter Cixi's dream of continuing to improve the functions of the Summer Palace and restore the scenery during the period before the reconstruction of *Qingyiyuan*. Besides continuing the unfinished project of "Taking good care of health", Cixi ordered Lei Tingchang to continue to finish the Office of Household Ministry, the imperial kitchen in the south of the Office of Labour Ministry and other subsidiary architectural buildings, and the design work of the important decorative architectural buildings in the southwest of the lake, including *Zhijing* Pavilion, *Changguan* Hall, and *Zaojian* Hall. In the eighth month of 1898, the 24th year of Guangxu's reign, the *Wuxu* Reform Movement irritated Cixi so strongly that she left Summer Palace for 19 months. In 1900, the 26th year of Guangxu's reign, the Eight-Power Allied Forces attacked Beijing, and Cixi escaped towards the west in panic. In 1902, the 28th year of Guangxu's reign, Cixi returned. Because of the difficult conditions, she had no interest in construction. Although the architect Lei Tingchang had already completed the design and model for the reconstruction of *Zhijing* Pavilion (Fig.22, Fig.23), the reconstruction was not implemented (Fig.24).

After the reconstruction in the late period of the Qing Dynasty, the Summer Palace has continued the general pattern of *Qingyiyuan* until now. However, to ensure that the living quarters and governance needs were met amidst the austerity Summer Palace and the erstwhile *Qingyiyuan* did have specific differences in terms of the ambience of the garden,

根据样式房会同算房拟定的相关《做法钱粮底册》和《工料银两细册》，颐和园《工程清单》记载，这些『颐养工程』一直持续到光绪二十一年（1895年）。

大戏楼、佛香阁、谐趣园以及东宫门外大量配套建筑等。据颐和园《工程清单》记载，这些『颐养工程』一直持续到光绪二十一年（1895年）。

五、后续工程

光绪十八年（1892年），『颐养工程』尚未全部告竣，慈禧太后已开始短期驻跸颐和园。光绪二十一年（1895年）甲午海战的失利，并未打消慈禧太后继续完善颐和园功能需求和再现清漪园时期景观的梦想，除继续未竟的『颐养工程』外，慈禧太后命雷廷昌继续完成户部公所、工部公所及迤南大膳房等附属建筑，以及治镜阁、畅观堂、藻鉴堂等西南湖区重要点景建筑的设计工作。光绪二十四年（1898年）八月，戊戌政变，慈禧太后深受刺激，离开颐和园长达十九个月。光绪二十六年（1900年）八国联军攻入北京，慈禧太后仓皇西逃。光绪二十八年（1902年），慈禧太后回銮，时局艰难，也无心再兴园工，由雷廷昌已经完成复建设计方案和烫样的治镜阁（图二十二、图二十三）也未能付诸实施（图二十四）。

经过晚清重修后的颐和园，总体上延续了清漪园时期的格局，并传承至今。但为满足居住、理政的功能需要，同时受经济条件所限，在园林空间氛围、景观关系以及点景建筑处理上颐和园与清漪园也存在一定的差异。

图二十三　冶镜阁烫样（故宫博物院藏）

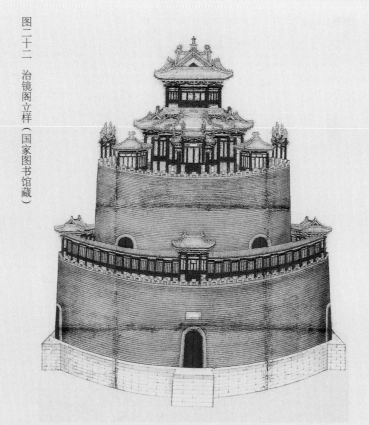

图二十二　冶镜阁立样（国家图书馆藏）

Fig.22　Elevation of *Zhijing* Pavilion (Collection of the National library)
Fig.23　Model of *Zhijing* Pavilion (Collection of the Palace Museum)

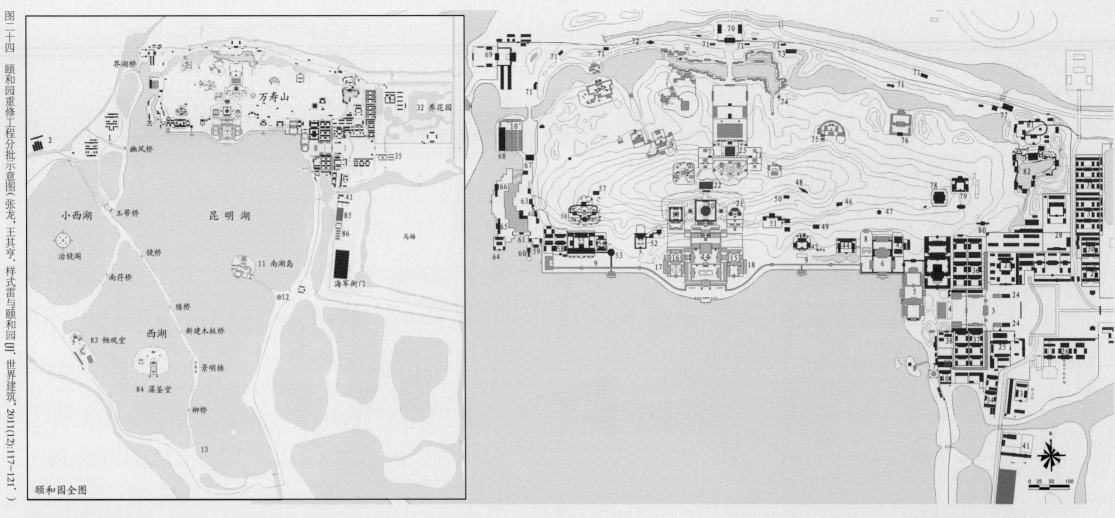

图二十四　颐和园重修工程分批示意图（张龙，王其亨．样式雷与颐和园[J]．世界建筑，2011(12):117—121.）

颐和园全图

■ 水操学堂相关 Engineering Projects of Naval Training School	17. 西九间 Nine-bay Room on the West	35. 内务府公所 Office of the Ministry of Internal Affairs	55. 听鹂馆 Tingli Pavilion	75. 多宝塔 Duobao Tower
1. 水操学堂 Naval Training School	18. 东九间 Nine-bay Room on the East	36. 寿膳房 Kitchen for the Empress Dowager	56. 画中游 Huazhongyou	76. 澹宁堂 Danning Hall
2. 西船坞 West Boatyard	■ 佛宇工程 Engineering Projects of Temple	37. 御膳房 Kitchen for the Emperor	57. 湖山真意 Hushanzhenyi	77. 眺远斋 Tiaoyuan Hall
■ 阅操工程 Engineering Projects for Review Training	19. 宝云阁 Baoyun Pavilion	38. 耶律楚材祠 Ancestral Temple of Yelv Chucai	58. 西四所 Four Courtyard Dwellings in the West	78. 景福阁 Jingfu Pavilion
3. 东宫门 East Palace Gate	20. 佛香阁 Foxiang Pavilion	39. 电灯公所 Office for Electric Lamp	59. 寄澜堂 Jilan Hall	79. 益寿堂 Yishou Hall
4. 仁寿殿 Renshou Hall	21. 转轮藏 Zhuanlunzang	40. 电灯公所迤南房间 Room on the South of Office for Electric Lamp	60. 石舫 Marble Boat	80. 紫气东来 Ziqidonglai Pavilion
5. 玉澜堂、宜芸馆 Yulan Hall and Yiyun Hall	22. 智慧海、众香界 Zhihuihai and Zhongxiangjie	41. 弹压处 Room for Patrolmen	61. 荇桥 Xing Bridge	81. 霁清轩 Jiqing Pavilion
6. 乐寿堂 Leshou Hall	23. 香岩宗印之阁 Xiangyanzongyin Pavilion	42. 德和园 Dehe Garden	62. 穿堂殿 Chuantang Hall	82. 谐趣园 Xiequ Garden
7. 文昌阁 Wenchang Gate Pavilion	■ 颐养工程 Engineering Projects for Supporting the Empress Dowager	43. 永寿斋 Yongshou Hall	63. 延清赏 Yanqingshang	■ 其他工程 Other Engineering Projects
8. 扬仁风 Yangrenfeng	24. 乾清侍卫值房 Room for Qianqing Guard	44. 养云轩 Yangyun Pavilion	64. 五圣祠 Wusheng Temple	83. 畅观堂 Changguan Hall
9. 长廊 Long Corridor	25. 銮仪卫 Room for Honor Guard	45. 无尽意轩 Wujinyi Pavilion	65. 迎旭楼 Yingxu Pavilion	84. 藻鉴堂 Zaojian Hall
10. 北船坞（两小）North Boatyard (two smaller ones)	26. 各项下处值房 Duty Room for a Series of Units	46. 福荫轩 Fuyin Pavilion	66. 澄怀阁 Chenghuai Pavilion	85. 工部公所 Office of the Ministry of Works
11. 南湖岛 Nanhu Island	27. 大他坦 Room for Entourage	47. 含新亭 Hanxin Pavilion	67. 宿云檐 Suyunyan Gate Pavilion	86. 大膳房 Big Kitchen
12. 阔如亭 Kuoru Pavilion	28. 南花园 South Flower Fostering Garden	48. 千峰彩翠 Qianfengcaicui Gate Pavilion	68. 北船坞（大船坞）North Boatyard (bigger one)	■ 弃建项目 Abandoned Projects
13. 西堤六桥花牌楼 Pailous of Six Bridges on the West Embankment	29. 步军统领衙门 Office of Public Security Bureau	49. 重翠亭 Chongcui Pavilion	69. 西宫门 West Palace Gate	
■ 正路工程 Engineering Projects of Central Building Complex	30. 堂档房 Archives	50. 意迟云在 Yichiyunzai	70. 北宫门 North Palace Gate	
14. 排云殿 Paiyun Hall	31. 升平署 Bureau of Drama	51. 写秋轩 Xieqiu Pavilion	71. 后山值房 Duty Room behind the hill	
15. 介寿堂 Jieshou Hall	32. 养花园 Flower Fostering Garden	52. 云松巢、邵窝 Yunsongchao and Shaowo	72. 通云 Tongyun Pavilion	
16. 清华轩 Qinghua Pavilion	33. 侍卫处 Room for Guard	53. 山色湖光共一楼 Shansehuguang-gongyilou Pavilion	73. 后溪河船坞 Back River Boatyard	
	34. 太医院、如意馆 Hospital and Room for Artist	54. 贵寿无极 Guishouwuji Hall	74. 寅辉城关 Yinhui Gate Pavilion	

说明：部分未核准建设年代和在1900年后修建的项目未列入本图，如乐农轩、草亭、外务部公所大堂等。

Instruction: Some projects without clear construction time and projects built after 1900 are not included in this drawing, such as Le Nong Pavilion, Cao Pavilion, and the lobby of the Office of the Ministry of Foreign Affairs.

Fig.24　Diagram of the Summer Palace reconstruction project in stages
(ZHANG Long, WANG qiheng. Yangshi Lei and the Summer Palace[J]. World Architecture, 2011(12):117—121.)

scenic relationships, and managing the decorative architectural buildings.

(1) The Ambience of the Garden

The open architectural pattern of the erstwhile *Qingyiyuan* and the arrangement of many religious buildings created a deeply religious ambience which facilitated religious activities like praying for the long life and happiness of the Empress Dowager. The Summer Palace created garden living space for taking good care of health by adding many buildings for living and naming them with the theme of "Happiness and Longevity", such as *Guishouwuji* Hall (Unlimited Fortune and Life), *JieshouHall* (Hall of Celebrating Longevity), *YongshouHall* (Hall of Eternal Life), and *YishouHall* (Hall of Increasing Longevity)…

(2) Scenic relationships

In the earlier period of *Qingyiyuan*, there were no walls on three sides except to the north of *Kunming* Lake. The scenic area extended outside through *Zhijing* Pavilion and the Farming and Weaving Picture Scenic Area in the west, and was connected with the surrounding scenes of paddy fields and gardens. The Qianlong Emperor actually managed *Sanshanwuyuan* (Three Mountains and Five Gardens) in the west as one holistic entity, so when he designed an architectural detail he took into account both the detail itself and its surroundings. The reconstruction of Summer Palace was limited by the poor state of the economy. The follies on the surface of western lake were almost all abandoned except for *Nanhu* Island. The Farming and Weaving Picture Scenic Area in the west was replaced by the Water Exercise School. Walls were added around *Kunming* Lake for safety. Because of the fading of other gardens, the whole scenery of *Sanshanwuyuan* disappeared. Thus Summer Palace will put more energy into the management of internal space and loose the connection with scenery around.

(3) Managing the Decorative of Architectural Buildings

The structure and shape of architectural buildings in the erstwhile *Qingyiyuan* were fashionable and heavily decorated as befitting the wealth, power and artistic culture of the Qianlong Emperor. However, when Summer Palace declined because of the shrinking economy and culture, many of the follies had to be downsized, the numbers of floors reduced and their decorations simplified.

（一）空间氛围

清漪园疏朗的建筑格局，众多宗教建筑的设置，营造浓郁的宗教山林气象，旨在通过礼佛、诵经等活动为太后延寿、祈福。而颐和园通过添建大量居住建筑并以「福寿」为题名，如贵寿无极、介寿堂、永寿斋、益寿堂……营造出适合颐养的园居空间。

（二）空间关系

清漪园时期昆明湖东、西、南三侧不设围墙，并通过西部的治镜阁、耕织图等景观建筑将清漪园的景域向外扩展，使之与周边稻田、园林景观连为一体。实际上乾隆皇帝是把西郊的三山五园作为一个整体来经营，在局部建筑空间处理时充分考虑与周边景观的借景关系。颐和园重修时囿于经济实力，除南湖岛外几乎完全放弃了西部湖面上的点景建筑，西部的耕织图也被水操学堂取代，出于安全考虑，昆明湖周围添建了围墙，再加上其他园林的衰败，三山五园的整体景观不复存在。因此，颐和园将更多的精力用于内部空间的经营，对于与周边景观的联系则比较放松。

（三）建筑造型

清漪园凭借着乾隆朝雄厚的财力和高度发达的文化艺术，其建筑结构、造型求新求异，装饰极尽繁复。而颐和园则因经济文化的衰落，诸多点景建筑不得不缩小规模、降低层数、简化装饰。

颐和园的文物建筑测绘

早在20世纪30年代，中国营造学社成员朱启钤、梁思成等人就提出，系统的测绘调查是解读、认识及传承中国建筑文化的基础工作。梁思成在《蓟县独乐寺观音阁山门考》中强调：

「我国古代建筑，澄之文献，所见颇多，周礼考工，阿房宫赋，两都两京，以至洛阳伽蓝记等等，固记载详尽，然吾侪所得，则隐约之印象……故研究古建筑，非作遗物实地调查测绘不可……结构之分析及制度之鉴别，在现状图之绘制。」

正是在这一认识的影响下，1934年，颐和园管理委员会完成了全园总图的测量（图二十五）。1935—1949年，在梁思成先生的指导下，北平文物整理委员会结合保养、维修工程，对荇桥、四大部洲、善现寺等建筑进行了简略测绘（图二十六）。

新中国成立后，清华大学、天津大学等高校结合教学与科研，程和档案建设，均对颐和园古建筑开展了较为系统的测绘工作，形成了丰硕的成果（图二十七）。

1998年，颐和园入选世界文化遗产，对其保护与传承提出了更高要求。鉴于古建筑测绘对完善颐和园档案建设，奠定文物保护基础，推动颐和园历史、文化、建筑、科学、艺术研究，彰显文化大国实力，促进公众对颐和园建筑艺术的理解等方面的重要作用。自2005年以来，北京市颐和园管理处联合天津大学，分别于2005年、2006年、2007年、2011年、2013年，投入五百余人次，借助三维激光扫描仪、GPS定位仪、摄影测量、无人机等先进的测量设备，对颐和园进行了系统

Survey and Recording of Heritage Buildings in Summer Palace

In the 1930's , Zhu Qiqian and Liang Sicheng, members of the Society for the Study of Chinese Architecture, pointed out that systematic survey and recording of heritage buildings was necessary and fundamental for reading, recognizing, and inheriting outstanding heritage buildings. Liang Sicheng stressed in the *Record of the Gate to Guanyin Pavilion of Dule Temple in Ji County*:

Ancient architectural buildings of China were recorded in literature, such as "Kaogongji" in *Zhouli*, "Prose-poem on E'Pang Palace" (*E'pang gong fu*), "Prose-poems on Two Capitals" (*Liang du fu*, by Ban Gu), "Prose-poems on Two Capitals" (*Er jing fu*, by Zhang Heng), *Records of Buddhist Temples in Luoyang* (*Luoyang Jialan ji*) and so on. Although the records are detailed, the information that we read are abstract expressions and in poetic language…Thus, survey and recording of heritage buildings is also necessary for research on ancient architectural buildings…The drawings of their current condition are helpful to analyze their structure and clarify which system of building regulations was being followed.

With this in mind, the Administrative Office of Summer Palace surveyed and measured the whole garden in 1934 (Fig.25). Between 1935 and 1949, under the instruction of Liang Sicheng, the Heritage Reorganization Committee of Beijing carried out simply surveys and measurements of the *Xing* Bridge, *Sidabuzhou*, *Shanxian* Temple and other architecture in connection with maintenance and repair work (Fig.26).

After the founding of the People's Republic of China, Tsinghua University, Tianjin University and other colleges and universities carried out survey and measurement of the Summer Palace in combination with both teaching and research. The Administrative Office of Summer Palace also implemented survey and measurement as the basis of their conservation work and the development of their archive. Together they all systematically surveyed and studied the ancient architectural buildings in Summer Palace and produced considerable outcomes (Fig.27).

In 1998, Summer Palace was selected for inclusion in the World Cultural Heritage list. Thus the requirements for heritage conservation have increased. One may consider the

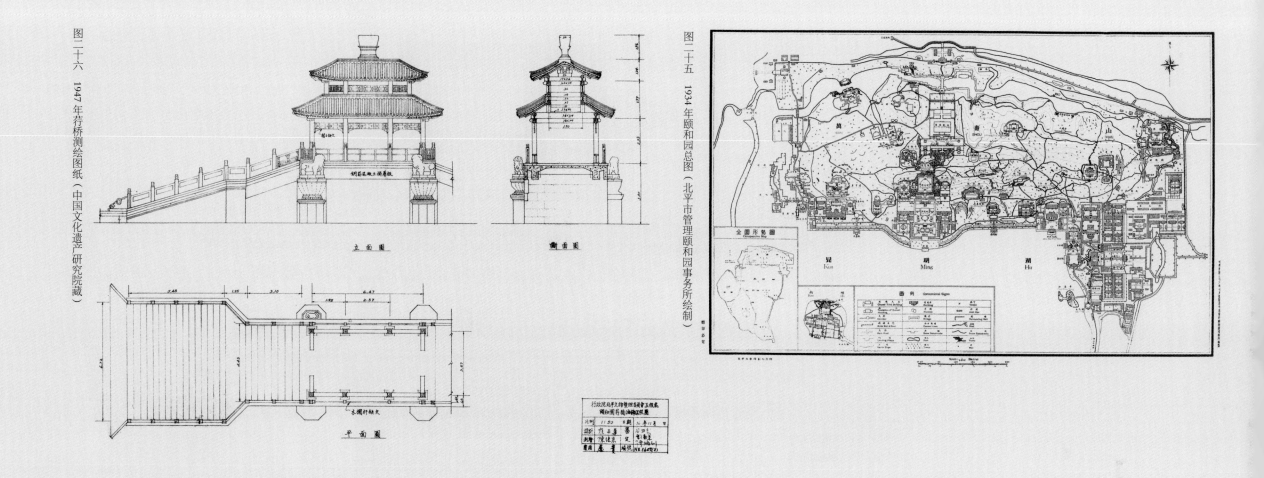

Fig.25　Master Plan of the Summer Palace in 1934 (Drawn by Administration Office of the Summer Palace of Beiping)

Fig.26　The Drawings of *Xing* Bridge in 1947 (Collection of Chinese Academy of cultural heritage)

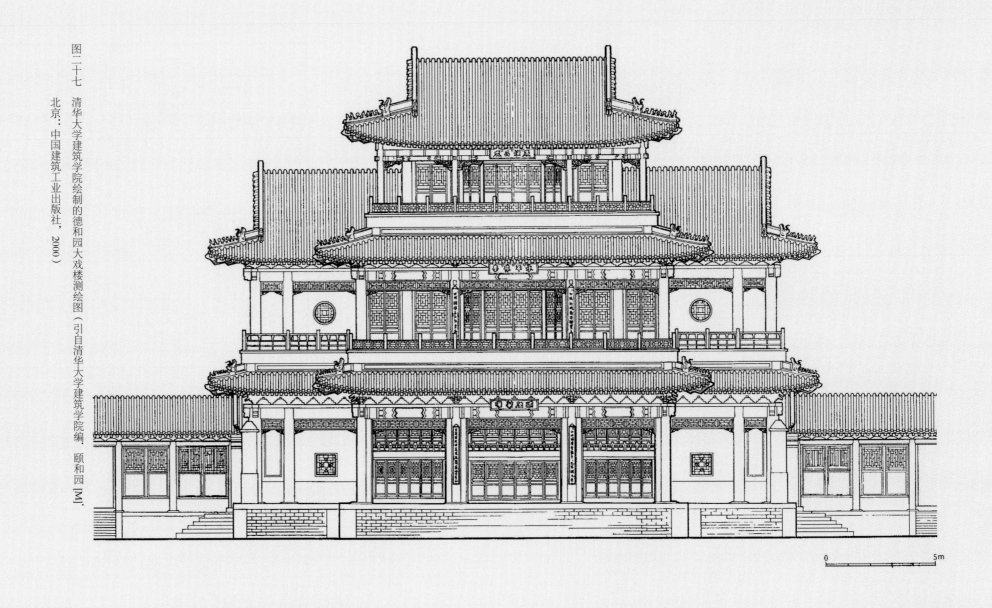

图二十七　清华大学建筑学院绘制的德和园大戏楼测绘图（引自清华大学建筑学院编．颐和园 [M]．

北京：中国建筑工业出版社，2000）

Fig.27　The Drawings of the Great Opera Hall of *Dehe* Garden by the School of Architecture, Tsinghua University
(Architecture School of Tsinghua University. *Summer Palace*[M]. Bejing: China Architecture & Building Press, 2000)

important effects that survey and measurement of ancient Chinese architecture has had on developing the archives of the Summer Palace: laying the foundations for heritage conservation; promoting research on the history, culture, architecture, science and art of Summer Palace; putting the soft power of Chinese culture on display; encouraging public understanding of the architecture of the Summer Palace. Since 2005, with the strong support of the Administrative Office of the Summer Palace, Tianjin University has deployed over 500 people together with 3D laser scanners, Global Positioning System (GPS), Stereographs, Drones or Unmanned Aerial Vehicles and other advanced measurement equipment to survey and record the Summer Palace digitally (Fig.28-Fig.31) and to draw over 3,000 drawings in 2005, 2006, 2007, 2011, and 2013. Over 90% of the work has now been finished. This work has elevated the archival development of traditional Chinese Architecture at the Summer Palace; laid solid foundations for Chinese architectural research on Summer Palace, and provided a wonderful databank for conserving China's architectural heritage.

The drawings published at this time have been chosen from all the survey and measurement drawings completed by Tianjin University since 1950. According to their architectural functions and images of scenery, these drawings can be divided into 13 types, including: Palace gates, palace, buddhist temples, daoist temples, city gate pavilions, residential architecture, theatrical architecture, street architecture, rural architecture, garden within gardens, bridges, landscape architecture displays. Through all these drawings we have completed since 1950. We wish to demonstrate the imperial garden tradition that Summer Palace the emperor's garden is a miniature of both the cosmos of heaven and earth and the emperor's heart with their shared ability to assimilate myriad different cultures.

的数字化测绘工作（图二十八至图三十一），绘制图纸 3000 余幅，完成全园 90% 以上古建筑的测绘工作。将颐和园古建筑档案建设推向新的阶段，为颐和园的相关研究奠定了坚实基础，也为相关文物保护工程提供了极大便利。

此次出版的图纸，就是从天津大学 1950 年代至今所测绘的图纸中精选而来，按建筑功能与景观意象分为宫门、宫殿、佛寺、祠庙、城关、居住建筑、观演建筑、街肆建筑、乡村建筑、园中园、桥、点景建筑、小品，总计 13 个类型，来进行编排，意在体现中国古代帝王『移天缩地在君怀』的园林营造传统，以及其对多元文化兼收并蓄的胸怀。

图二十八　宝云阁角亭屋面测量（作者自摄）

图三十一　利用无人机对德和园建筑群进行信息采集（作者自摄）

图三十　利用三维激光扫描仪进行德和园建筑群的扫描（作者自摄）

图二十九　排云殿屋面测量（作者自摄）

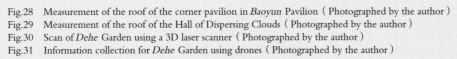

Fig.28　Measurement of the roof of the corner pavilion in *Baoyun* Pavilion（Photographed by the author）
Fig.29　Measurement of the roof of the Hall of Dispersing Clouds（Photographed by the author）
Fig.30　Scan of *Dehe* Garden using a 3D laser scanner（Photographed by the author）
Fig.31　Information collection for *Dehe* Garden using drones（Photographed by the author）

注　释

① 黄成彦．颐和园昆明湖3500余年沉积物研究[M]．北京：海洋出版社，1996：3．

② 元代周伯琦《仲秋休沐日同崇文僚佐游西山纪事二首》．近光集．卷三．文渊阁《四库全书》内联网版．

③ [清]于敏中，等．钦定日下旧闻考[M]．北京：北京古籍出版社，1983：1411．

④ [清]于敏中，等．钦定日下旧闻考[M]．北京：北京古籍出版社，1983：1408．

⑤ 乔宇《白岩集》．引自：钦定日下旧闻考．卷八十五．文渊阁《四库全书》内联网版．

⑥ 宋启明《长安可游记》．引自：钦定日下旧闻考．卷八十四．文渊阁《四库全书》内联网版．

⑦ 袁中道《珂雪斋集》．引自：万寿山昆明湖记．文渊阁《四库全书》内联网版．

⑧ 钦定日下旧闻考．卷八十四．万寿山昆明湖记．文渊阁《四库全书》内联网版．

⑨ 御制文集·初集卷四．麦庄桥记．文渊阁《四库全书》内联网版．

⑩ 乾隆九年（1744年）其在《圆明园后记》中曾写道："天宝地灵之区，帝王豫游之地，无以逾此……并明白昭告后世子孙必不舍此而重费民力以创建苑囿，斯则深契朕发皇考勤俭之心以为心矣。"

⑪ 『昆明湖上旧有龙神祠，爰新葺之，而名之曰：广润。』御制诗集·二集卷十七．文渊阁《四库全书》内联网版．

⑫ 御制诗集·二集卷十八．董邦达西湖图．文渊阁《四库全书》内联网版．

⑬ 御制诗集·三集卷九十八．小西冷．『西峰浸水西湖似，缀景西冷小肖诸。何必孤山忆风景，已看仲夏淀芙蕖。』文渊阁《四库全书》内联网版．

⑭ 御制诗集·二集卷三十八．万寿山即事．文渊阁《四库全书》内联网版．

⑮ 御制诗集·初集卷二十二．月地云居．文渊阁《四库全书》内联网版．

⑯ 御制诗集·二集卷八十七．海岳开襟歌．文渊阁《四库全书》内联网版．

⑰ 钦定日下旧闻考·卷八十四．御制万寿山清漪园记．文渊阁《四库全书》内联网版．

⑱ 奏销档．中国第一历史档案馆藏．

⑲ 中国第一历史档案馆．内务府档案．奉宸苑第4602号，转引：叶志如，唐益年．光绪朝的三海工程与北洋海军——兼论颐和园工程挪用北洋海军经费问题，明清档案与历史研究学术讨论会论文，第12号．

⑳ 张嘉懿先生藏同治十三年正月十四日《内务府拆用旧木植及请旨采办木料奏底》引自：刘敦桢，刘敦桢文集（一）[M]．北京：中国建筑工业出版社，1982：355．

㉑ 中国第一历史档案馆．陈设册．档号：1995．

㉒ 中国第一历史档案馆．陈设册．档号：5193．

㉓ 中国第一历史档案馆．陈设册．档号：6256．

㉔ 中国第一历史档案馆．陈设册．档号：6252．

㉕ 中国第一历史档案馆．陈设册．档号：6265．

㉖ 中国第一历史档案馆．陈设册．档号：6370．

㉗ 中国第一历史档案馆．陈设册．档号：3861．

㉘ 中国第一历史档案馆．陈设册．档号：779．

㉙ 中国第一历史档案馆．录副奏折．档号：03-2808-022．

㉚ 中国第一历史档案馆．录副奏折．档号：03-2808-022．

㉛ 中国第一历史档案馆．录副奏折．档号：03-3646-069．

㉜ 张侠，杨治本，等．清末海军史料[M]．北京：海洋出版社，1982：919．

㉝ 王道成．颐和园修建年代考[J]．北京市论文集，1980：257-261．

㉞ [清]世续，等．清实录·德宗实录（四）[M]．北京：中华书局，1987（55）：529．

㉟ 藏于中国第一历史档案馆，《工程清单》是对颐和园工程进展情况的记录，恭备慈览，每五天一记，始于光绪十七年一月，截止光绪二十一年五月。

㊱ [清]世续，等．清实录·德宗实录（四）[M]．北京：中华书局，1987（55）：591．

Notes

① Huang Chengyan. *Study on the Sediments for Over 3500 Years in the Kunming Lake of the Summer Palace*[M]. Beijing: China Ocean Press, 1996: 3.

② Zhou Boqi. *Two poems recording a visit to the West Hills with Chongwen colleagues on a holiday in mid autumn.* Jinguang Ji. vol.3. *Sikuquanshu* [Intranet].

③ Yu Minzhong et al. *Qindingrixiajiuwenkao*[M]. Beijing: Beijing Gu ji chu ban she, 1983:1411.

④ Qiao Yu. *Baiyanji*, From: [Qing] YU Minzhong etc. *Textual Criticism on Old News in the Capital*[M]. Beijing: Beijing Ancient Books Press, 1983: 1408.

⑤ Song Qiming. *Chang'an ke you ji*, in *Qindingrixiajiuwenkao*. vol.84. *Sikuquanshu* [Intranet].

⑥ Yuan Zhongdao, *Kexuezhaiji*, in *Qindingrixiajiuwenkao*. vol. 85. *Sikuquanshu* [Intranet].

⑦ *Qindingrixiajiuwenkao*.Vol. 84. Record of Kunming Lake and Longevity Hill. *Sikuquanshu* [Intranet].

⑧ "Record of Maizhuang Bridge," Book1, vol. 4. *Yuzhiwenji*. *Sikuquanshu* [Intranet].

⑨ In the 9th year of Qianlong's reign (1744), he wrote in the *Later record of Yuanmingyuan*: "This area enjoys the blessing of heaven and earth. An emperor cannot have a better place for enjoyment than this … my descendants should not abandon this garden and construct more gardens. Thus [you] will appreciate me for following my father's desire for austerity."

⑩ "The original Dragon King Temple on *Kunming* Lake was repaired recently and renamed Guangrun." *Yuzhishiji*, Book 2, vol. 17. *Sikuquanshu* [Intranet].

⑪ "Drawing of West Lake by Dong Bangda", *Yuzhishiji*. Book 2, vol.18. *Sikuquanshu* [Intranet].

⑫ "On Longevity Hill", *Yuzhishiji*. Book 2, vol.38, *Sikuquanshu* [Intranet]

⑬ "West Hill in water, this scenery is like the West Lake; the decorative scenes were designed to imitate *Xiling* area. There is no need to recall the scenery of Gu Hill. Plentiful lotus appear here in summer." In *Little Xiling*, *Yuzhishiji*, Book3, vol.98, *Sikuquanshu* [Intranet].

⑭ "Record of Qingyiyuan Longevity Hill", *Qindingrixiajiuwenkao*, vol.84. *Sikuquanshu* [Intranet].

⑮ "Accounting Reports of the Inner Court," in The First Historical Archives of China.

⑯ "Haiyuekaijinge", *Yuzhishiji*, Book 2, vol.87, *Sikuquanshu* [Intranet].

⑰ "Haiyuekaijinge", *Yuzhishiji*, Book 1, vol.22, *Sikuquanshu* [Intranet].

⑱ The First Historical Archives of China. Furnishings Volume.No.1995.

⑲ The First Historical Archives of China. Furnishings Volume.No.5193.

⑳ The First Historical Archives of China. Furnishings Volume.No.6256.

㉑ The First Historical Archives of China. Furnishings Volume.No.6252.

㉒ The First Historical Archives of China. Furnishings Volume.No.6265.

㉓ The First Historical Archives of China. Furnishings Volume.No.6370.

㉔ The First Historical Archives of China. Furnishings Volume.No.3861.

㉕ The First Historical Archives of China. Furnishings Volume.No.779.

㉖ The First Historical Archives of China. Lufuzouzhe.No.03-2808-022.

㉗ The First Historical Archives of China. Lufuzouzhe.No.03-2808-022.

㉘ The First Historical Archives of China. Lufuzouzhe.No. 03-3646-069.

㉙ *Record of Removing and Using Old Timbers and proposal for purchasing timber by the Ministry of Internal Affairs*, dated January 14th in the 13th year of Tongzhi(lunar calendar), collected by ZHANG Jiayi. From:LIU Dunzhen. *Collected Works of LIU Dunzhen(vol.1)*[M]. Beijing:China Architecture and Building Press,1982:355.

㉚ The First Historical Archives of China. Archive of the Inner Court. Fengchenyuan, No.4602. From:Ye Zhiru and Tang Yinian. *Sanhai Construction and Beiyang Navy in Guangxu's Reign: Discussion about the embezzlement of navy funds for the Summer Palace project.* Proceedings of the *Ming-Qing Archives and History Study Conference*. No.12.

㉛ Zhang Xia. Yang Zhiben etc. *The history of the navy in the late period of Qing*[M]. Beijing: China Ocean Press, 1982:919.

㉜ WANG Daocheng. *Study on the Construction Time of the Summer Palace*[J]. *Collection of Beijing Historical Essays*,1980:257-261.

㉝ [Qing]SHI Xu etc. *Qing Veritable Records·Emperor Dezong(4)*[M].Beijing:China Press,1987(55): 529.

㉞ Collected in the First Historical Archives of China. *The construction List is records of the development of the Summer Palace.* It was written up every five days for Cixi. Start in January 1891 (the 17th year of the reign of Guangxu's reign) and finishing in May of 1895 (the 21st year of Guangxu's reign).

㉟ WANG Daocheng. *Study on the Construction Tme of the Summer Palace*[J]. *Collection of Beijing Historical Essays*, 1980: 257-261.

㊱ [Qing]SHI Xu etc. *Qing Veritable Records·Emperor Dezong(4)*[M]. Beijing: China Press, 1987(55): 591.

图

版

Drawings

宫门
Palace Gate

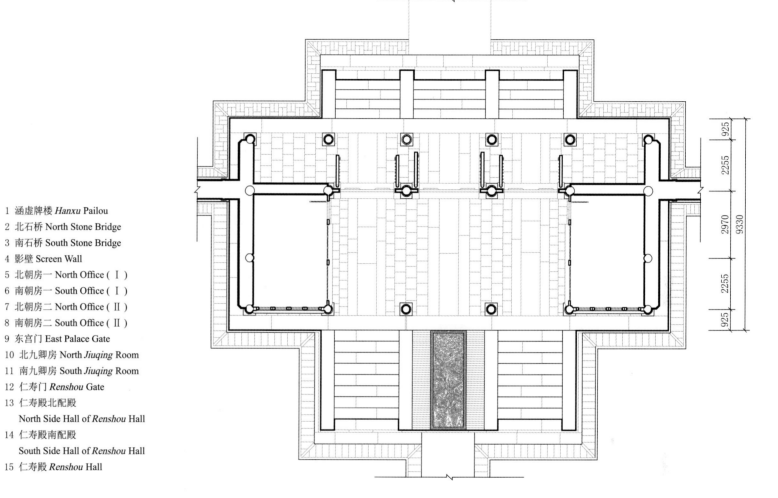

1　涵虚牌楼 *Hanxu* Pailou
2　北石桥 North Stone Bridge
3　南石桥 South Stone Bridge
4　影壁 Screen Wall
5　北朝房一 North Office（Ⅰ）
6　南朝房一 South Office（Ⅰ）
7　北朝房二 North Office（Ⅱ）
8　南朝房二 South Office（Ⅱ）
9　东宫门 East Palace Gate
10　北九卿房 North *Jiuqing* Room
11　南九卿房 South *Jiuqing* Room
12　仁寿门 *Renshou* Gate
13　仁寿殿北配殿
　　North Side Hall of *Renshou* Hall
14　仁寿殿南配殿
　　South Side Hall of *Renshou* Hall
15　仁寿殿 *Renshou* Hall

注：东宫门建筑群是门殿一体式布局，其中仁寿殿图纸见宫殿部分第 54 至 58 页。

Note: The architectural complex of the East Palace Gate is an integrated layout of the gate and palace, the drawings of the *Renshou* Hall can be found on pages 54 to 58 of the palace section.

东宫门组群总平面图
Site Plan of the East Palace Gate Complex

0　4　　20　　40m

东宫门平面图
Plan of the East Palace Gate

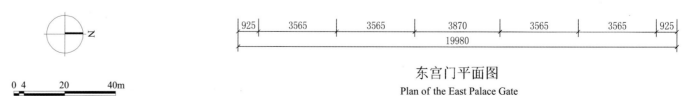

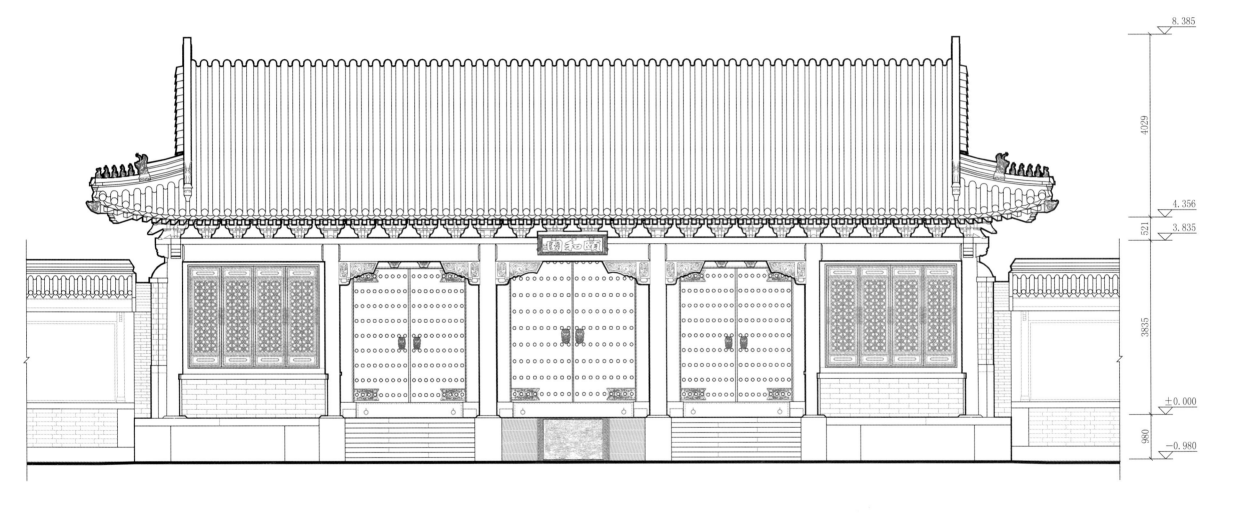

8.385

4029

4.356

521

3.835

3835

±0.000

980

−0.980

925 3565 3565 3870 3565 3565 925

19980

东宫门正立面图

Front Elevation of the East Palace Gate

8.385

1505

6.880

950

5.930

765

5.165

692

4.473

638

3.835

3835

±0.000

955

−0.955

925　　2255　　　　　　5225　　　925

9330

东宫门明间剖面图

Mingjian Section of the East Palace Gate

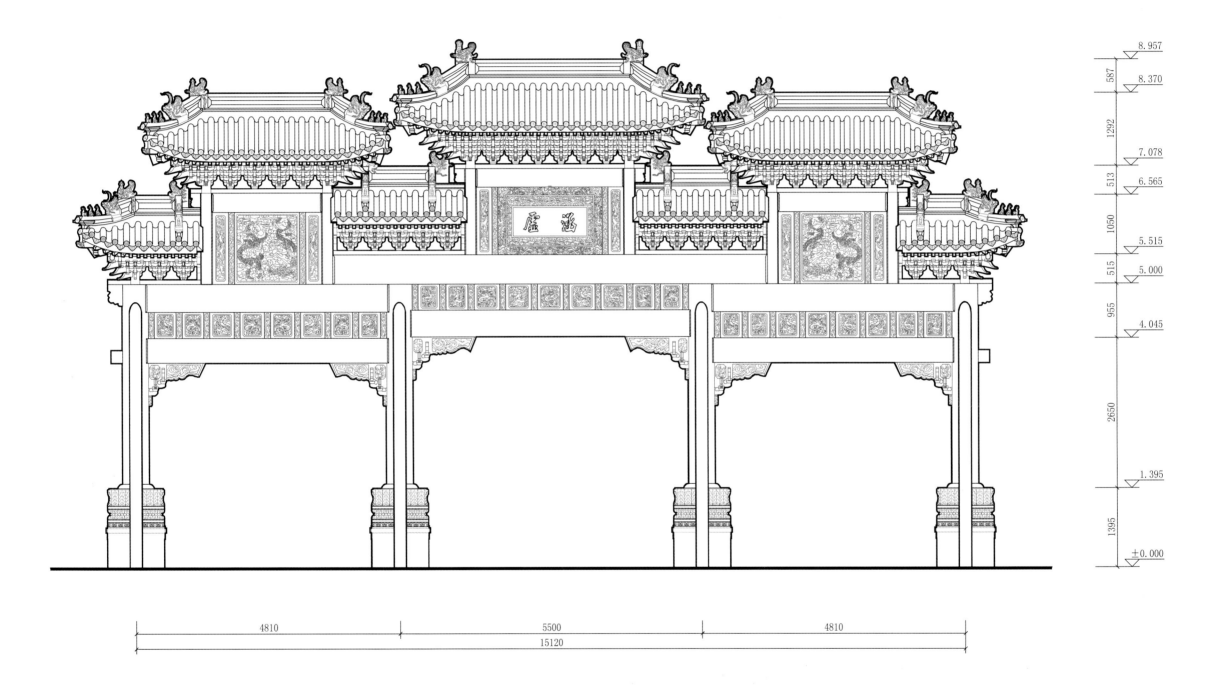

涵虚牌楼正立面图
Front Elevation of the *Hanxu* Pailou

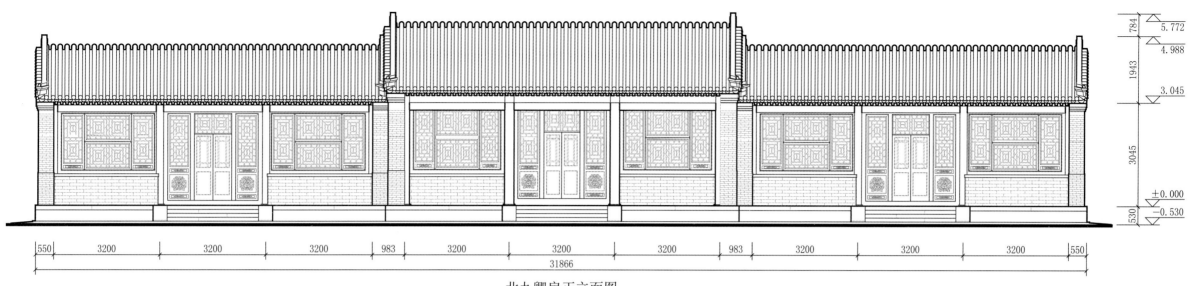

北九卿房正立面图

Front Elevation of the North *Jiuqing* Room

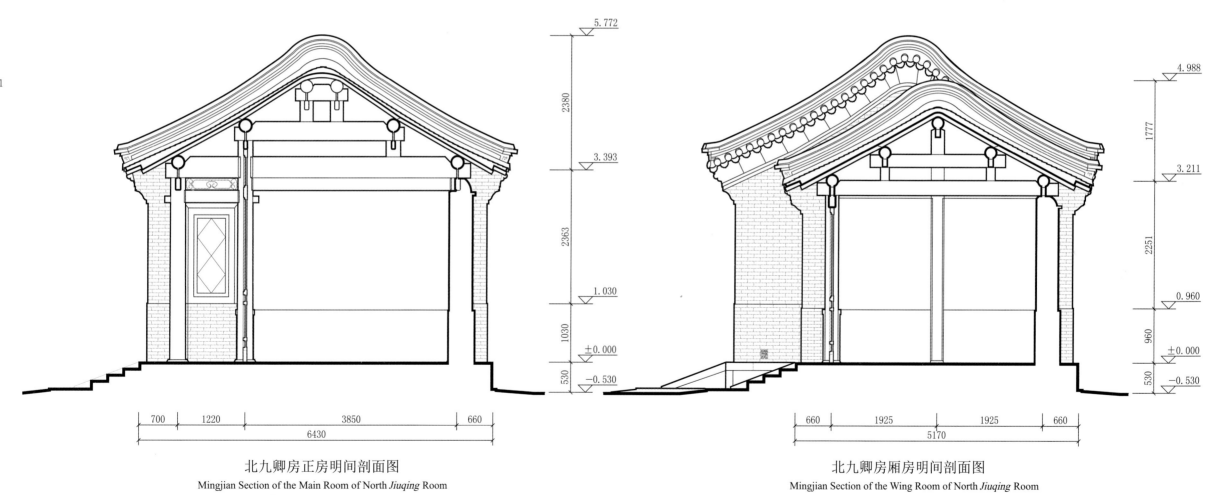

北九卿房正房明间剖面图

Mingjian Section of the Main Room of North *Jiuqing* Room

北九卿房厢房明间剖面图

Mingjian Section of the Wing Room of North *Jiuqing* Room

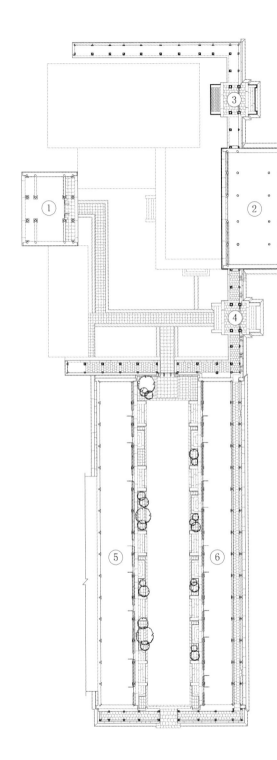

1 西宫门 West Palace Gate
2 德兴殿 *Dexing* Hall
3 德兴殿北垂花门
  North Festooned Door of *Dexing* Hall
4 德兴殿南垂花门
  South Festooned Door of *Dexing* Hall
5 德兴殿西值房
  West Duty Room of *Dexing* Hall
6 德兴殿东值房
  East Duty Room of *Dexing* Hall

N

西宫门组群总平面图
Site Plan of the West Palace Gate Complex

0 1   5   10m

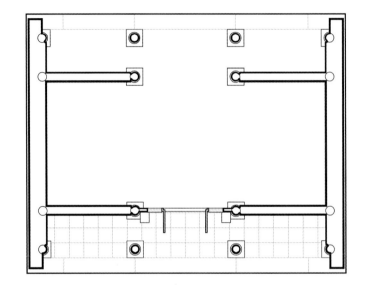

西宫门平面图
Plan of the West Palace Gate

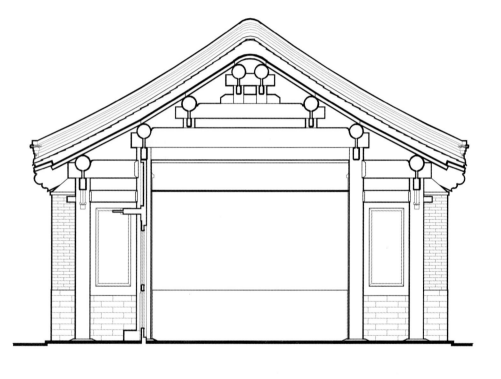

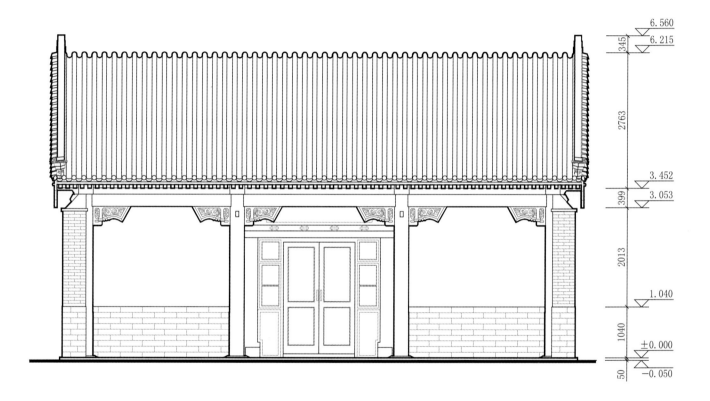

西宫门明间剖面图
Mingjian Section of the West Palace Gate

西宫门正立面图
Front Elevation of the West Palace Gate

8.234

365   7.869

3848

4.021

2989

1.032

1032

±0.000

735   −0.735

505   3132   3346   3820   3346   3132   505

17786

德兴殿正立面图

Front Elevation of the *Dexing* Hall

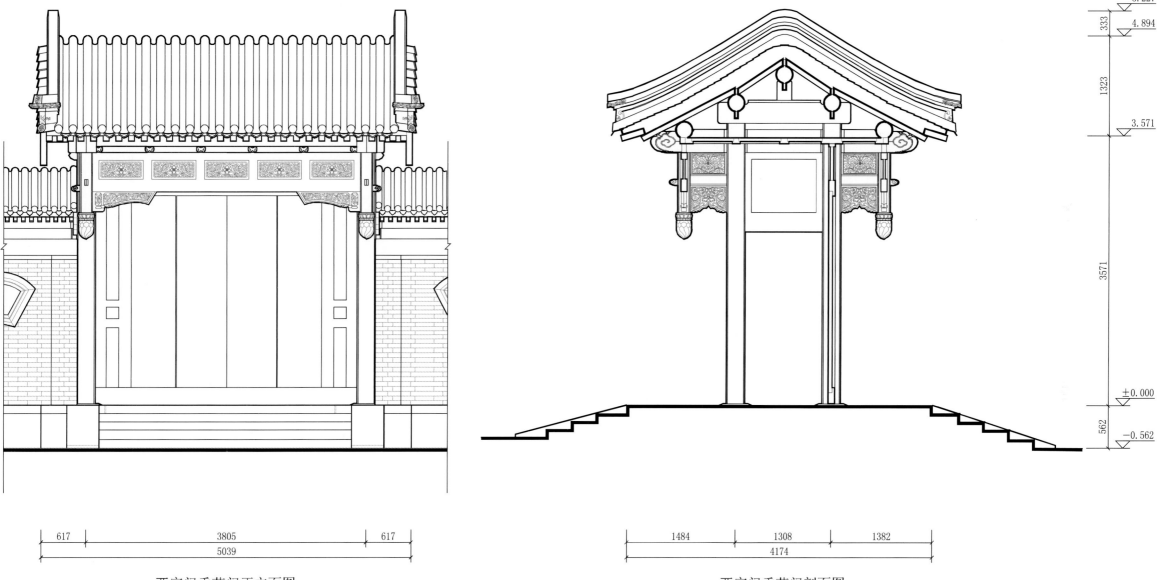

5.227

333

4.894

1323

3.571

3571

±0.000

562

−0.562

617　3805　617

5039

1484　1308　1382

4174

西宫门垂花门正立面图
Front Elevation of the Festooned Door of the West Palace Gate

西宫门垂花门剖面图
Section of the Festooned Door of the West Palace Gate

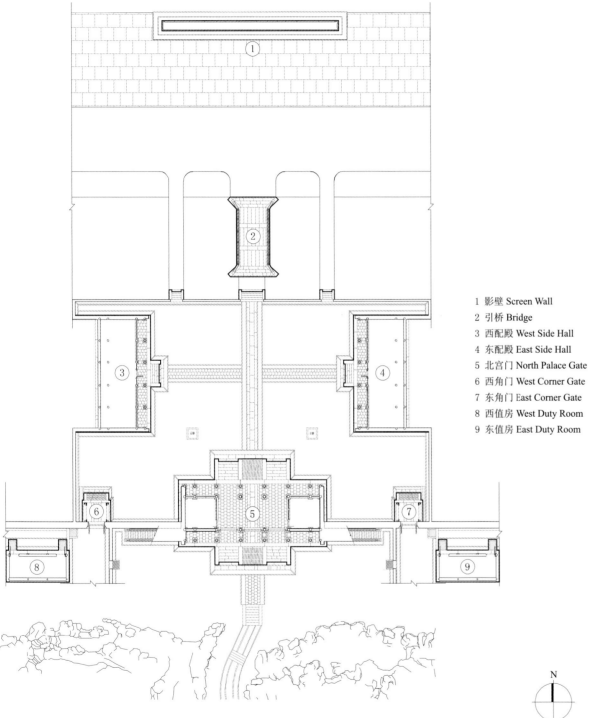

1　影壁　Screen Wall
2　引桥　Bridge
3　西配殿　West Side Hall
4　东配殿　East Side Hall
5　北宫门　North Palace Gate
6　西角门　West Corner Gate
7　东角门　East Corner Gate
8　西值房　West Duty Room
9　东值房　East Duty Room

北宫门组群总平面图
Site Plan of the North Palace Gate Complex

0 1　　5　　10m

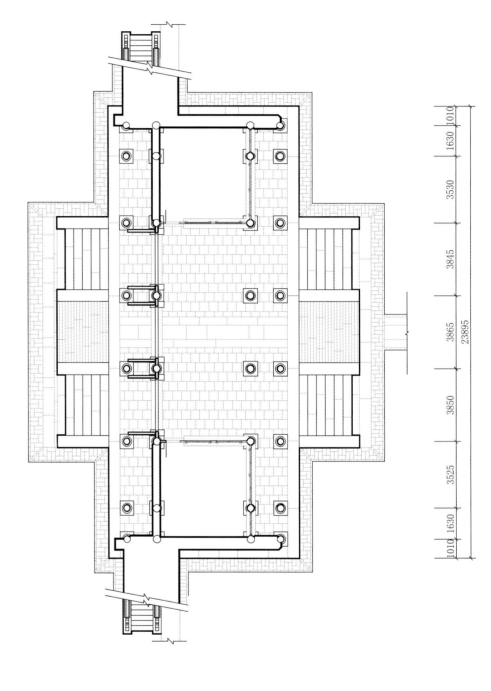

北宫门平面图
Plan of the North Palace Gate

北宫门组群总立面图

Elevation of the North Palace Gate Complex

0　1　　　　　5m

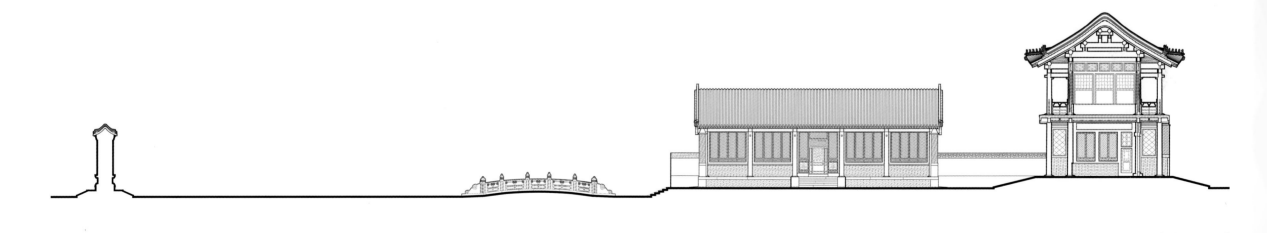

北宫门组群总剖面图

Section of the North Palace Gate Complex

0　1　　　　　5m

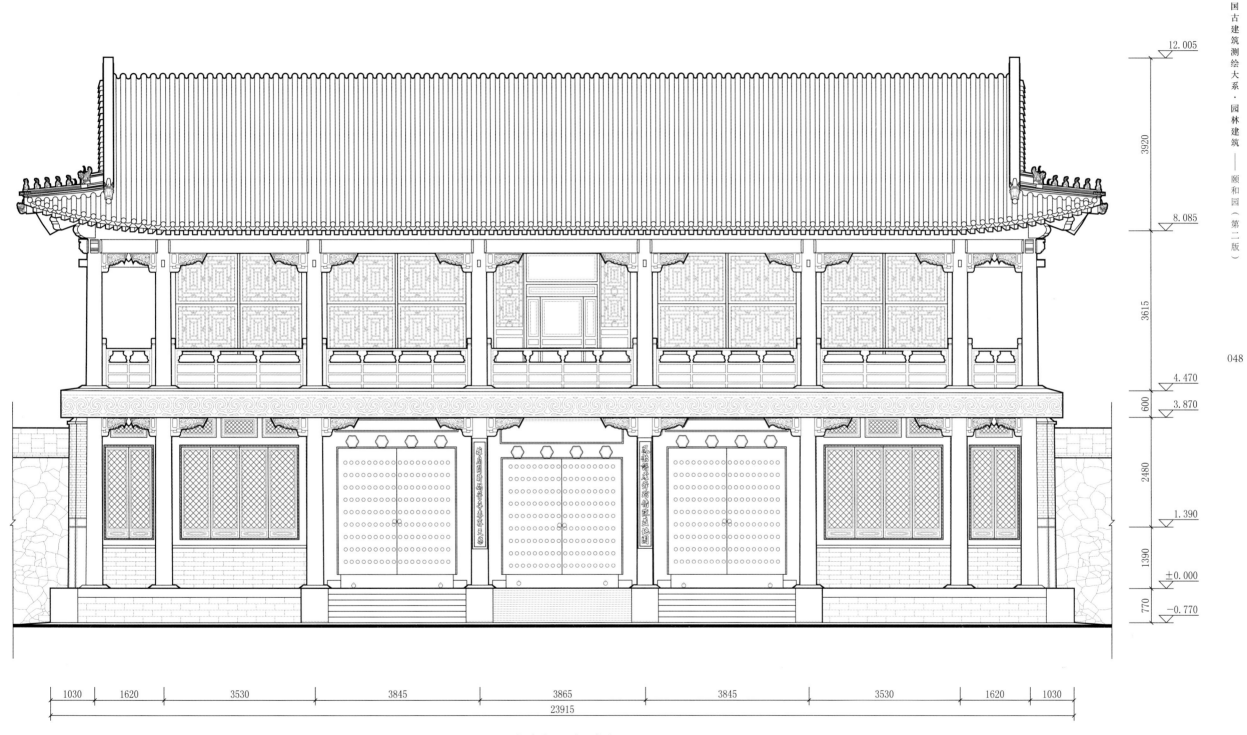

12.005

3920

8.085

3615

4.470

600　3.870

2480

1.390

1390

±0.000

770　−0.770

| 1030 | 1620 | 3530 | 3845 | 3865 | 3845 | 3530 | 1620 | 1030 |

23915

北宫门正立面图

Front Elevation of the North Palace Gate

北宫门侧立面图
Side Elevation of the North Palace Gate

北宫门明间剖面图
Mingjian Section of the North Palace Gate

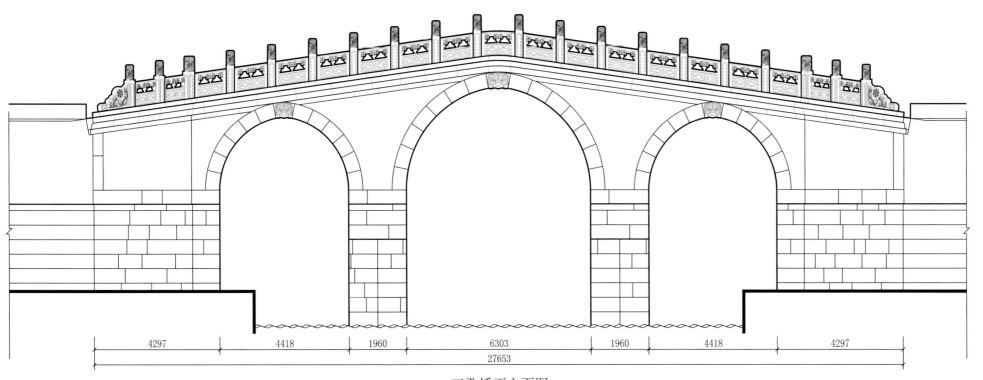

7.585
855
6.730
1125
5.605
3925
1.680
1680
±0.000

4297   4418   1960   6303   1960   4418   4297
27653

三孔桥正立面图
Front Elevation of the Three-arch Bridge

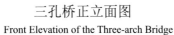

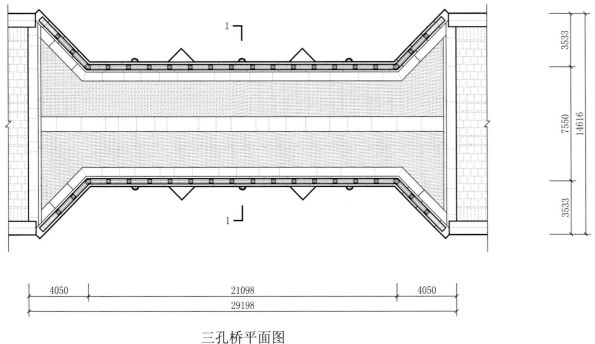

3533
7550   14616
3533

4050   21098   4050
29198

三孔桥平面图
Plan of the Three-arch Bridge

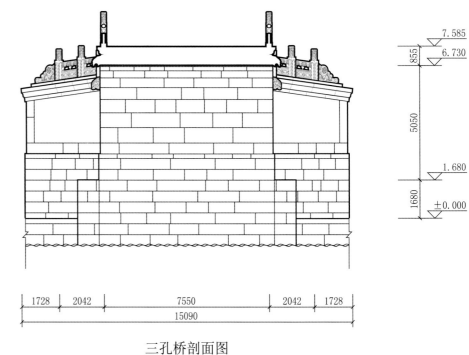

7.585
855
6.730

5050

1.680
1680
±0.000

1728   2042   7550   2042   1728
15090

三孔桥剖面图
Section of the Three-arch Bridge

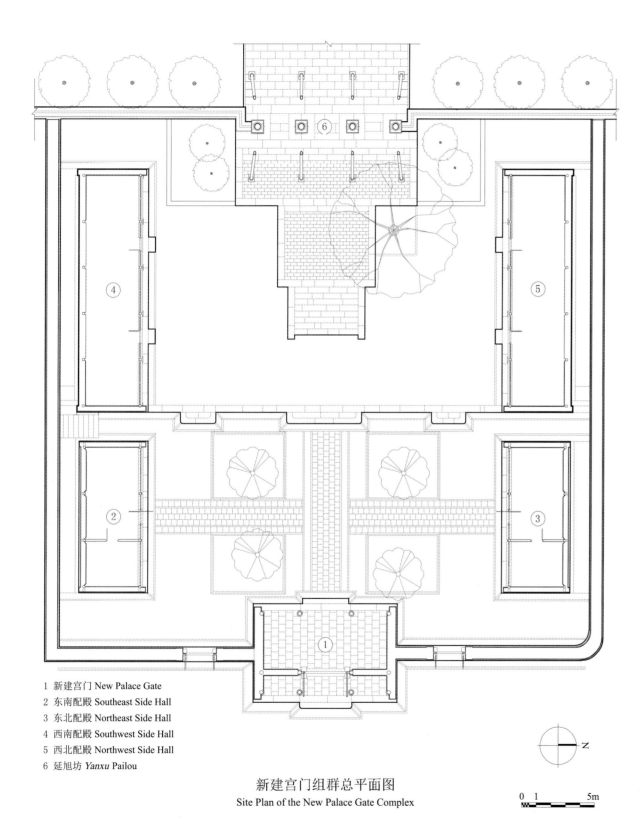

1 新建宫门 New Palace Gate
2 东南配殿 Southeast Side Hall
3 东北配殿 Northeast Side Hall
4 西南配殿 Southwest Side Hall
5 西北配殿 Northwest Side Hall
6 延旭坊 *Yanxu* Pailou

新建宫门组群总平面图
Site Plan of the New Palace Gate Complex

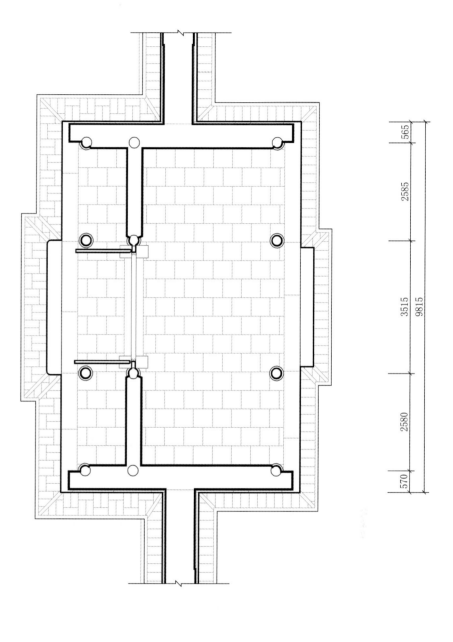

新建宫门平面图
Plan of the New Palace Gate

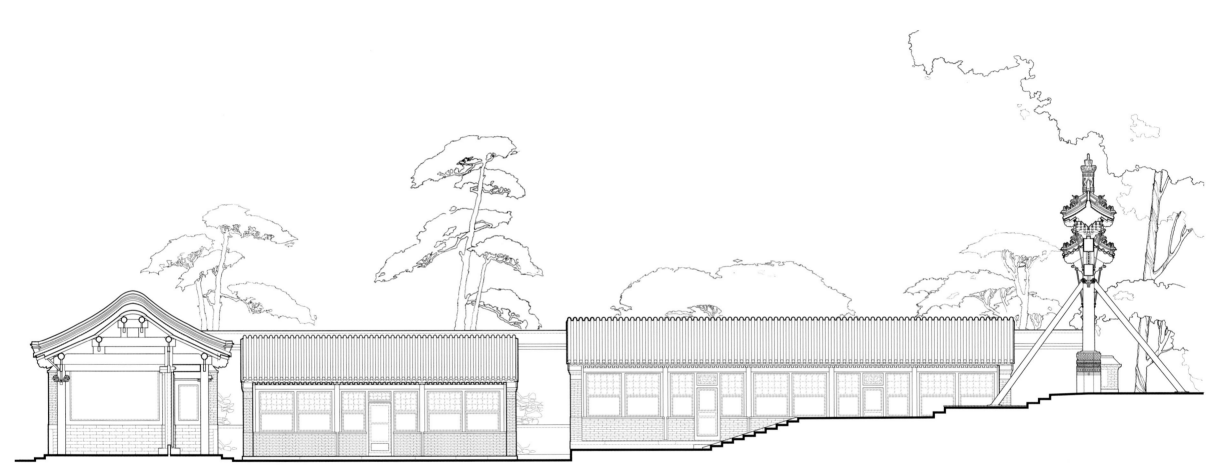

新建宫门组群总剖面图
Section of the New Palace Gate Complex

0 0.6 3m

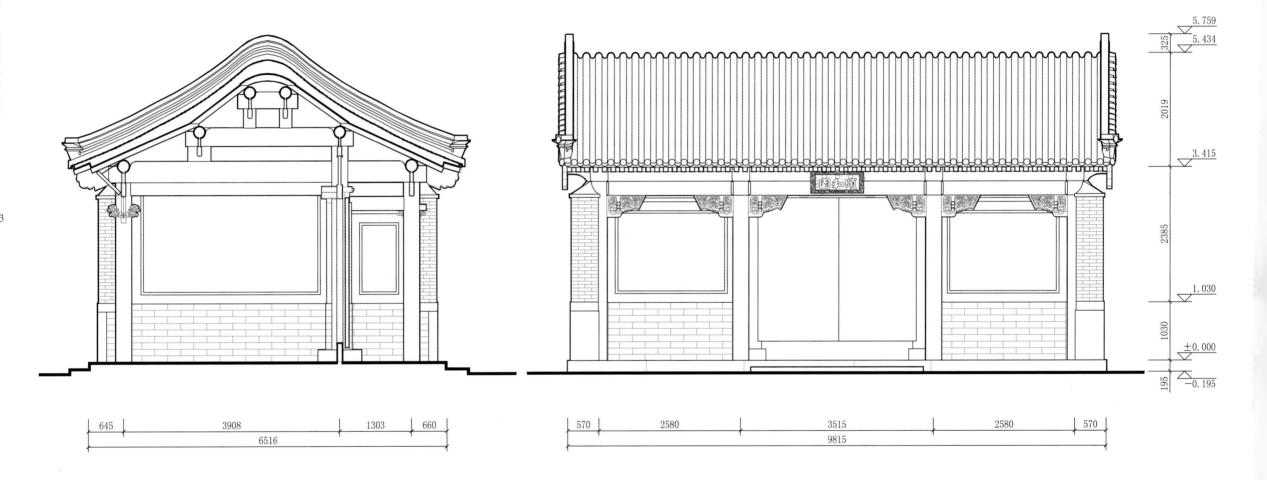

新建宫门明间剖面图
Mingjian Section of the New Palace Gate

新建宫门正立面图
Front Elevation of the New Palace Gate

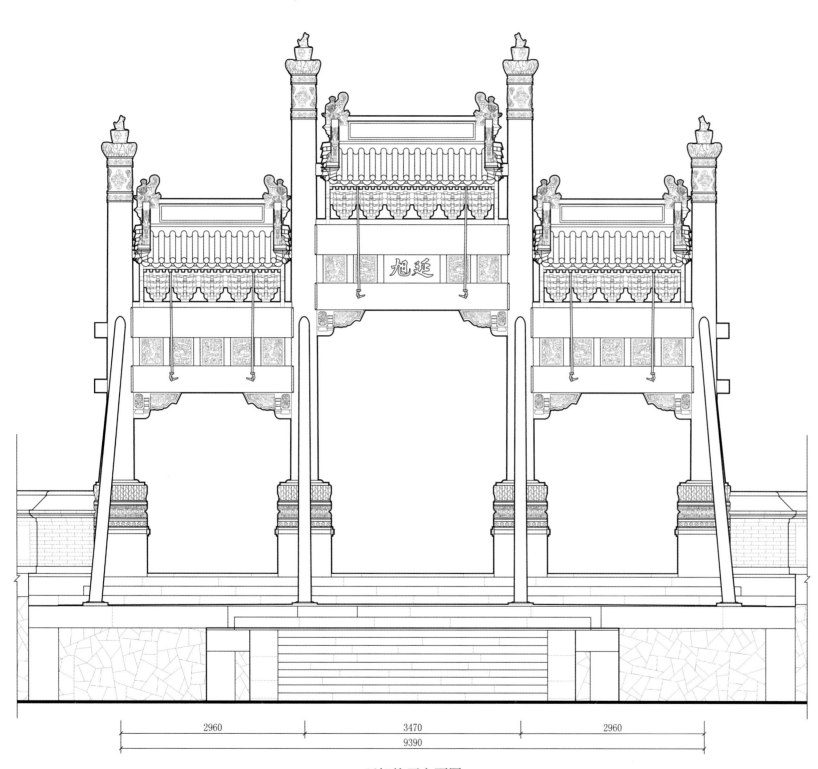

8.415

1290

7.125

1295

5.830

1130

4.700

700

4.000

2585

1.415

1415

±0.000

505

−0.505

370

−0.875

1105

−1.980

2960　　　3470　　　2960

9390

延旭坊正立面图

Front Elevation of the *Yanxu* Pailou

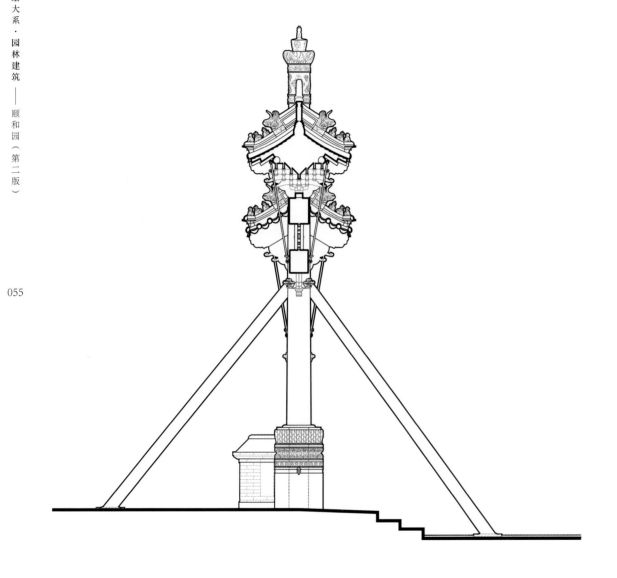

8.415

1295

7.120

1290

5.830

1130

4.700

700

4.000

2585

1.415

1415

±0.000

505

-0.505

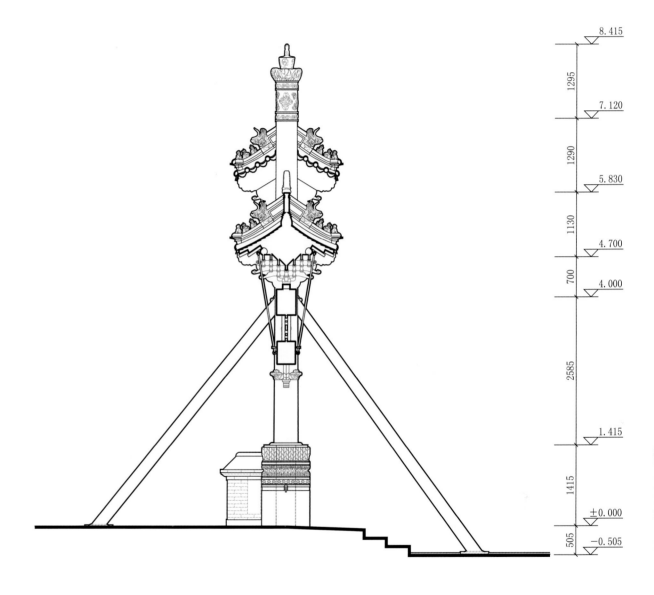

2940　865　2973

6778

延旭坊明间剖面图
Mingjian Section of the *Yanxu* Pailou

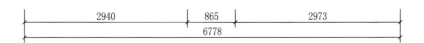

2940　865　2973

6778

延旭坊次间剖面图
Cijian Section of the *Yanxu* Pailou

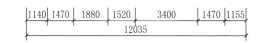

仁寿殿平面图
Plan of the *Renshou* Hall

仁寿殿北配殿平面图
Plan of the North Side Hall of *Renshou* Hall

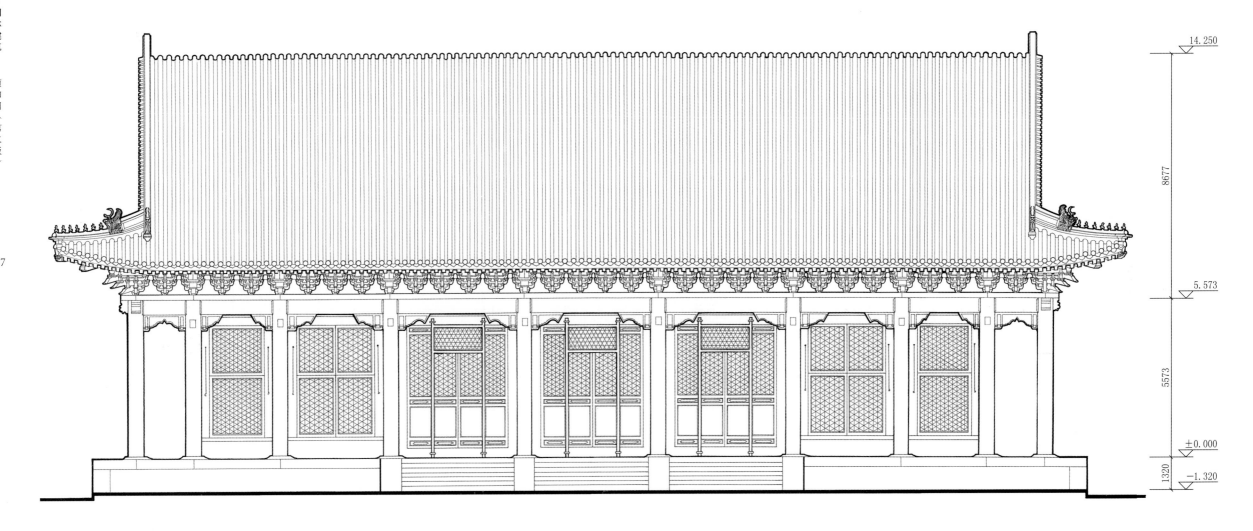

14.250

8677

5.573

5573

±0.000

1320

-1.320

1570　2090　3140　3965　4935　4920　4935　3965　3140　2090　1570

36320

仁寿殿正立面图

Front Elevation of the *Renshou* Hall

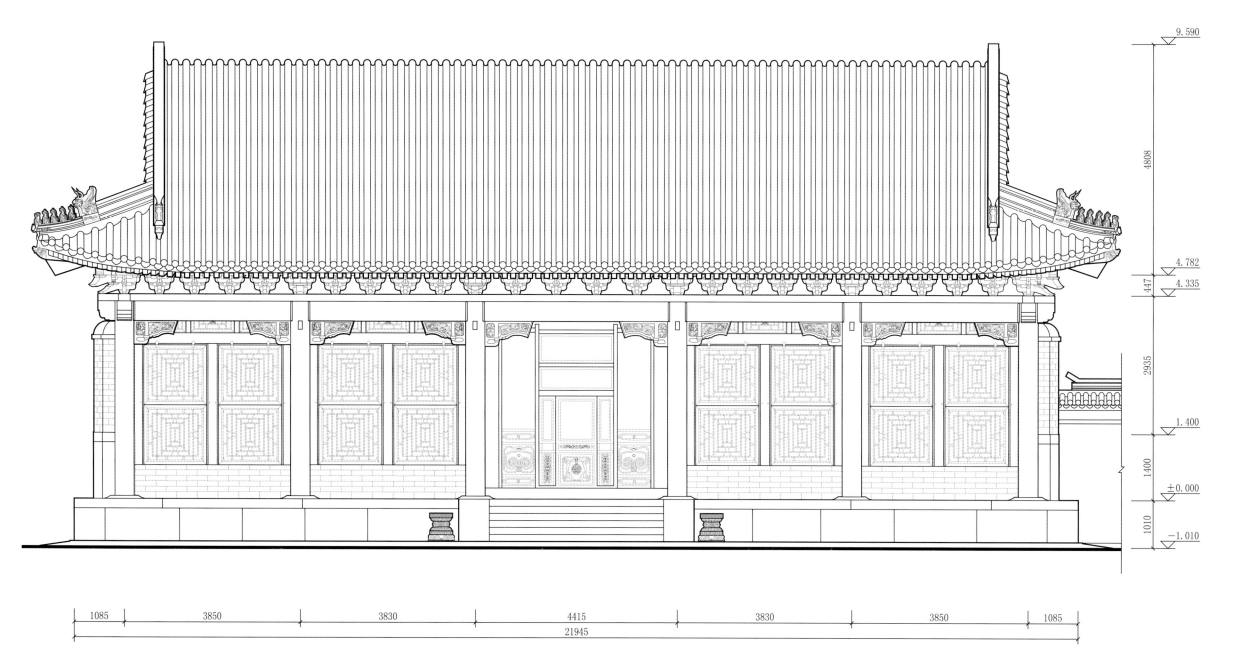

9.590

4808

4.782

447

4.335

2935

1.400

1400

±0.000

1010

−1.010

| 1085 | 3850 | 3830 | 4415 | 3830 | 3850 | 1085 |

21945

仁寿殿北配殿正立面图

Front Elevation of the North Side Hall of *Renshou* Hall

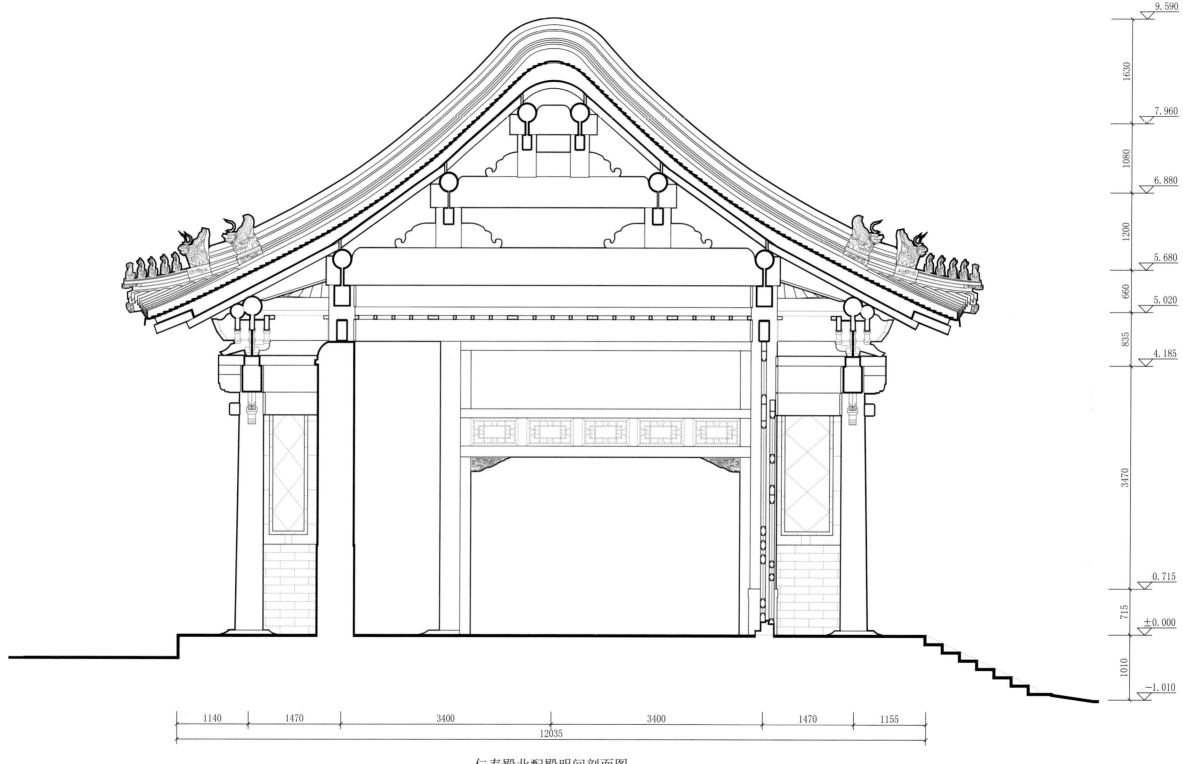

9.590

1630

7.960

1080

6.880

1200

5.680

660

5.020

835

4.185

3470

0.715

715

±0.000

1010

−1.010

1140 1470 3400 3400 1470 1155

12035

仁寿殿北配殿明间剖面图

Mingjian Section of the North Side Hall of *Renshou* Hall

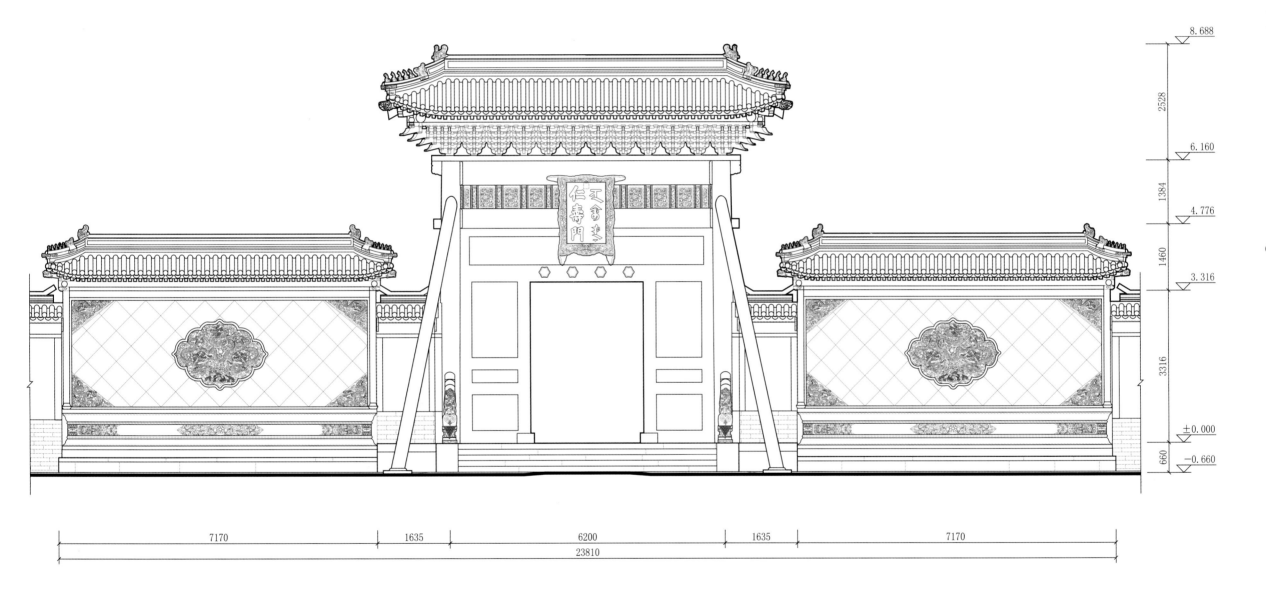

8.688

2528

6.160

1384

4.776

1460

3.316

3316

±0.000

660   −0.660

7170   1635   6200   1635   7170

23810

仁寿门正立面图

Front Elevation of the *Renshou* Gate

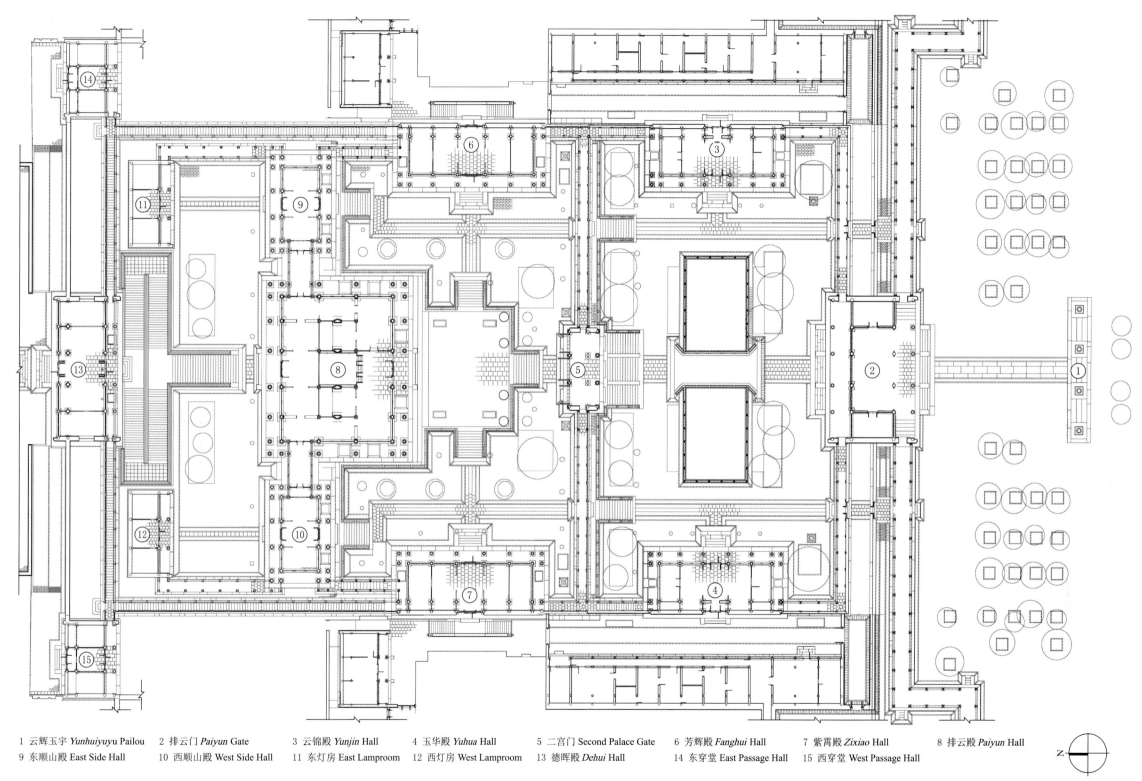

1 云辉玉宇 *Yunhuiyuyu* Pailou　　2 排云门 *Paiyun* Gate　　3 云锦殿 *Yunjin* Hall　　4 玉华殿 *Yuhua* Hall　　5 二宫门 Second Palace Gate　　6 芳辉殿 *Fanghui* Hall　　7 紫霄殿 *Zixiao* Hall　　8 排云殿 *Paiyun* Hall
9 东顺山殿 East Side Hall　　10 西顺山殿 West Side Hall　　11 东灯房 East Lamproom　　12 西灯房 West Lamproom　　13 德晖殿 *Dehui* Hall　　14 东穿堂 East Passage Hall　　15 西穿堂 West Passage Hall

排云殿组群总平面图
Site Plan of the *Paiyun* Hall Complex

0 1　　5　　10m

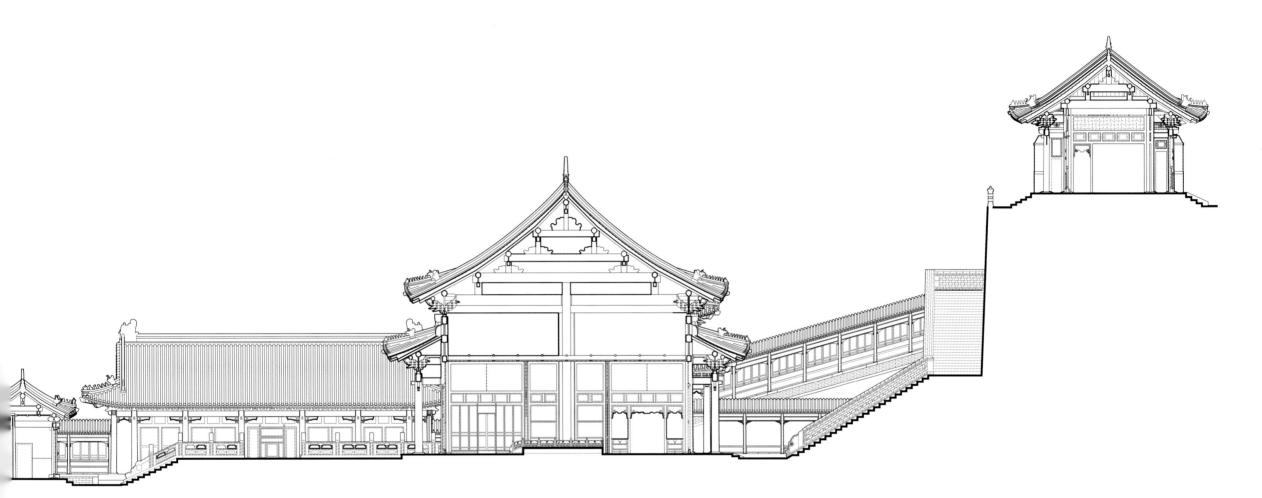

0 1   5m

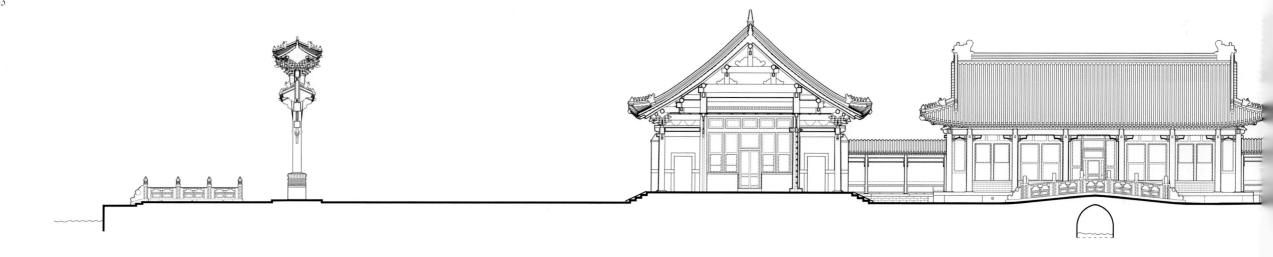

排云殿组群总剖面图

Section of the *Paiyun* Hall Complex

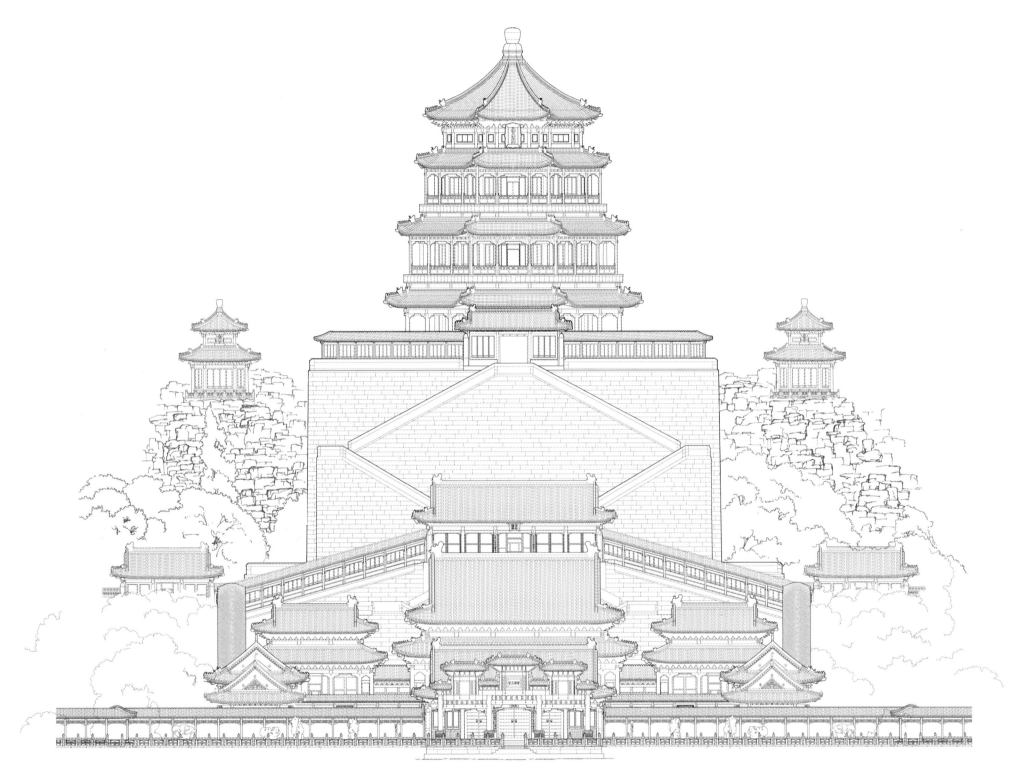

排云殿——佛香阁组群总立面图

Elevation of the *Paiyun* Hall and *Foxiang* Tower Complex

0 1　　5　　10m

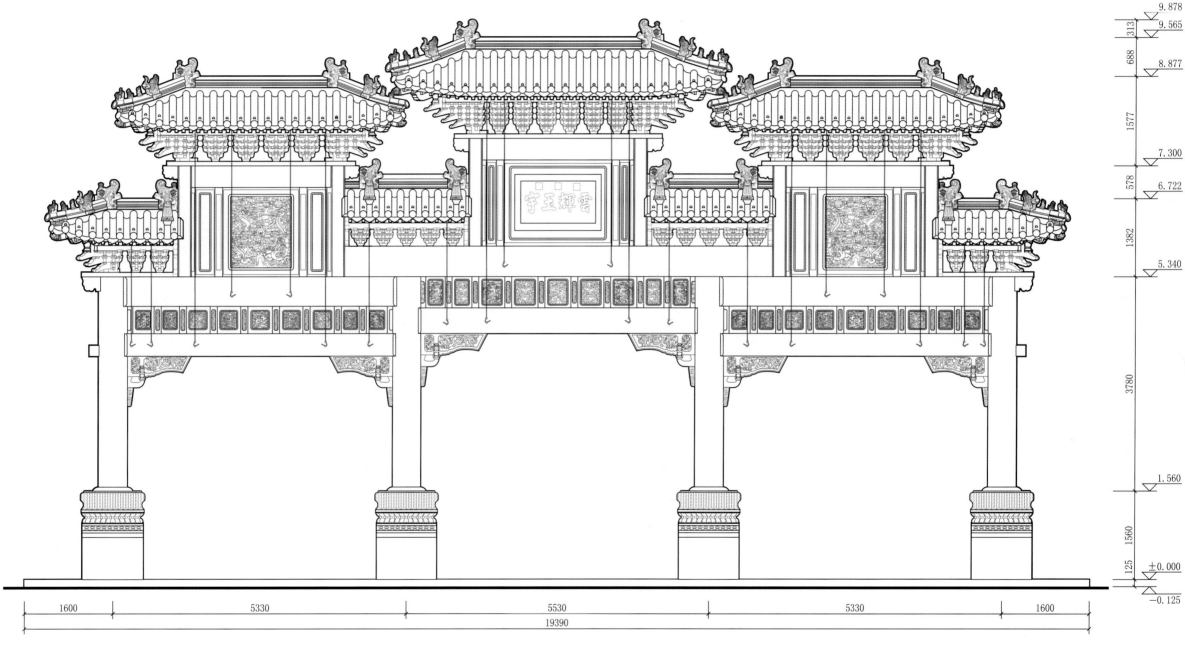

云辉玉宇牌楼正立面图

Front Elevation of the *Yunhuiyuyu* Pailou

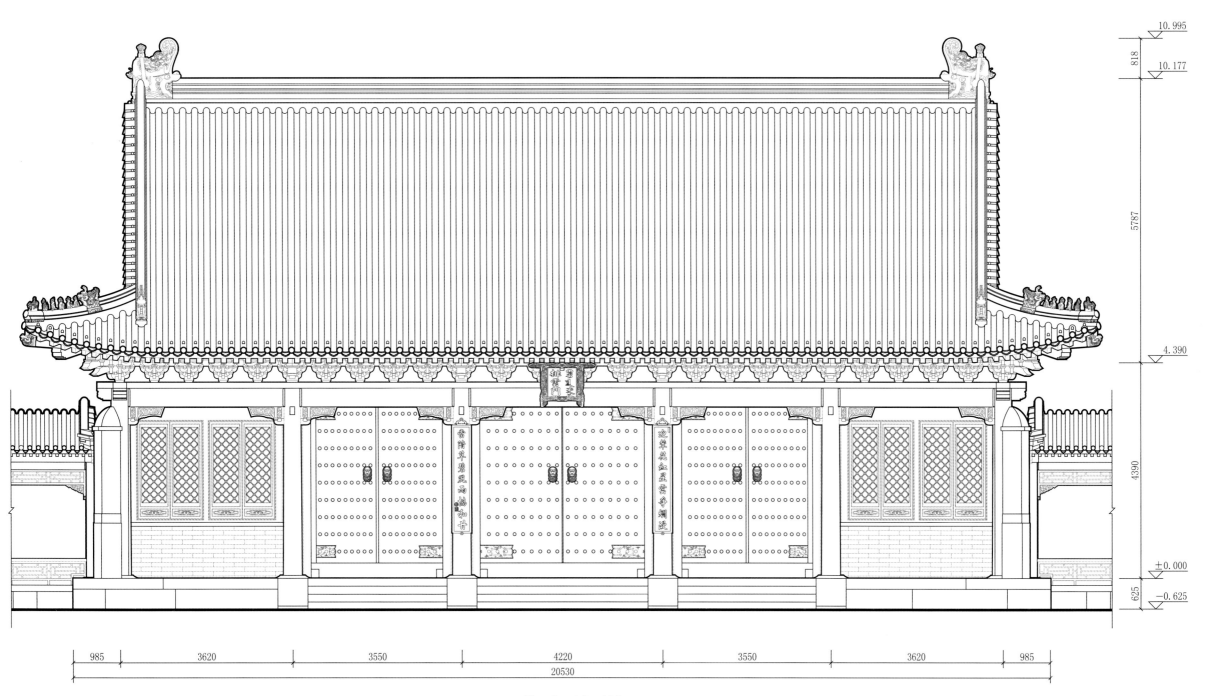

10.995

818

10.177

5787

4.390

4390

±0.000

625

−0.625

985　　3620　　3550　　4220　　3550　　3620　　985

20530

排云门正立面图
Front Elevation of the *Paiyun* Gate

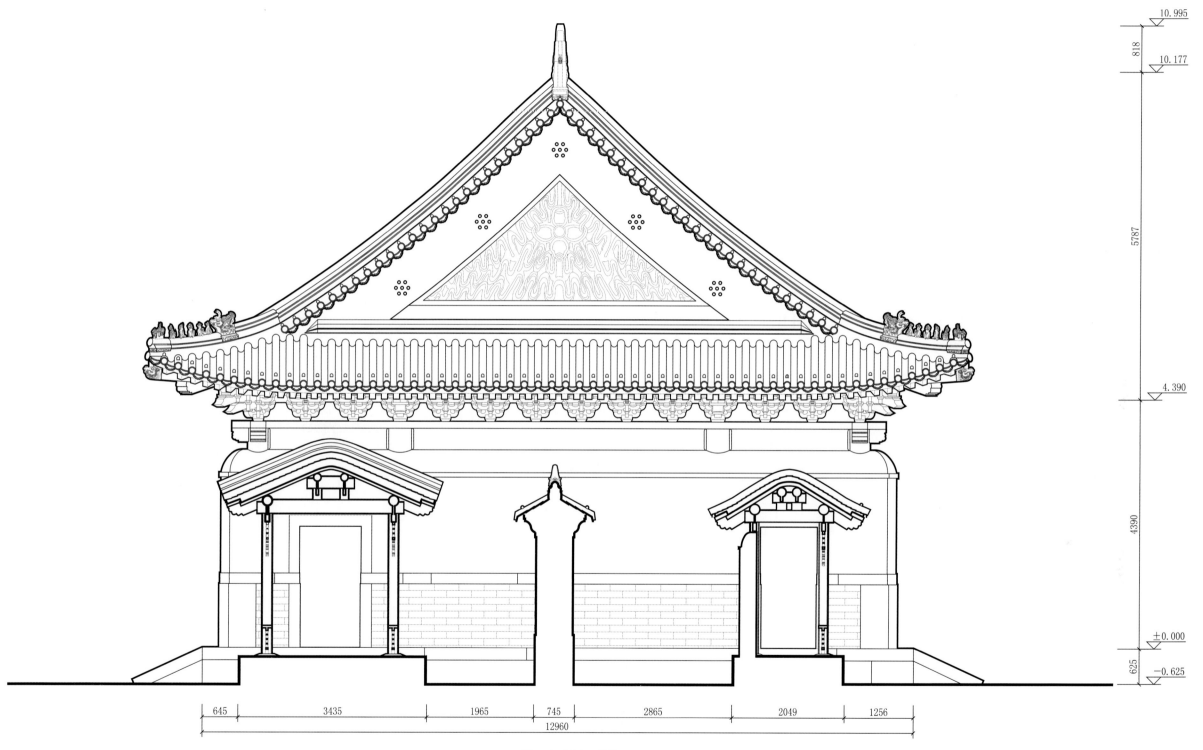

排云门侧立面图
Side Elevation of the *Paiyun* Gate

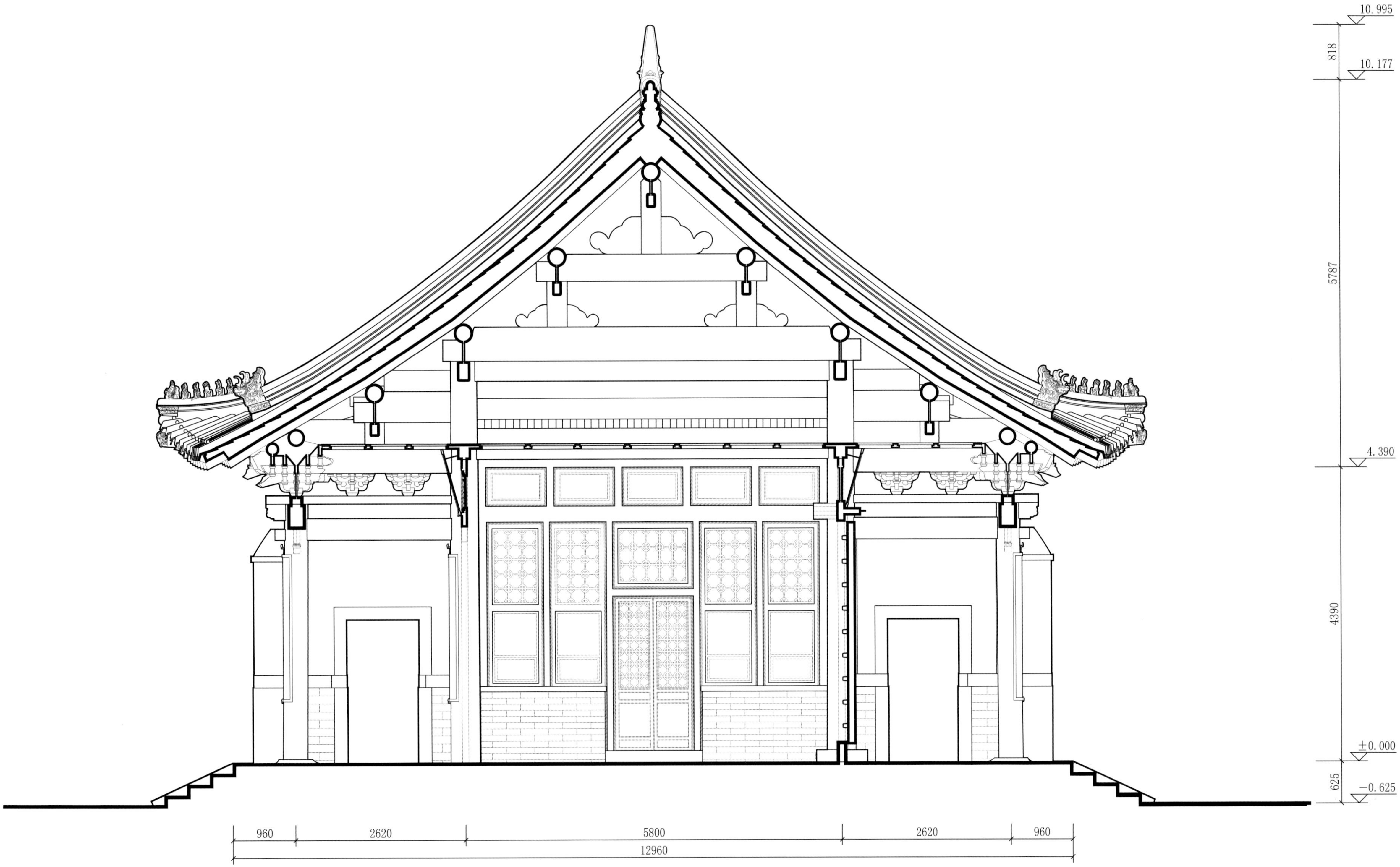

10.995

818

10.177

5787

4.390

4390

±0.000

625

−0.625

960　2620　5800　2620　960

12960

排云门明间剖面图

Mingjian Section of the *Paiyun* Hall

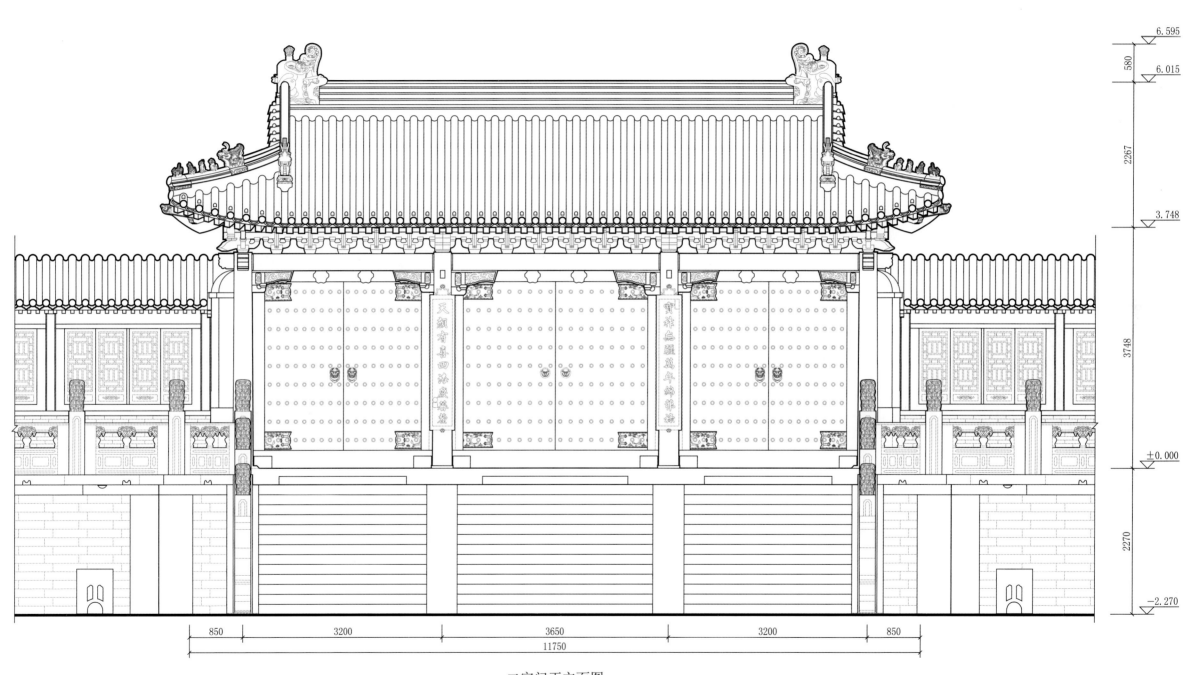

二宫门正立面图

Front Elevation of the Second Palace Gate

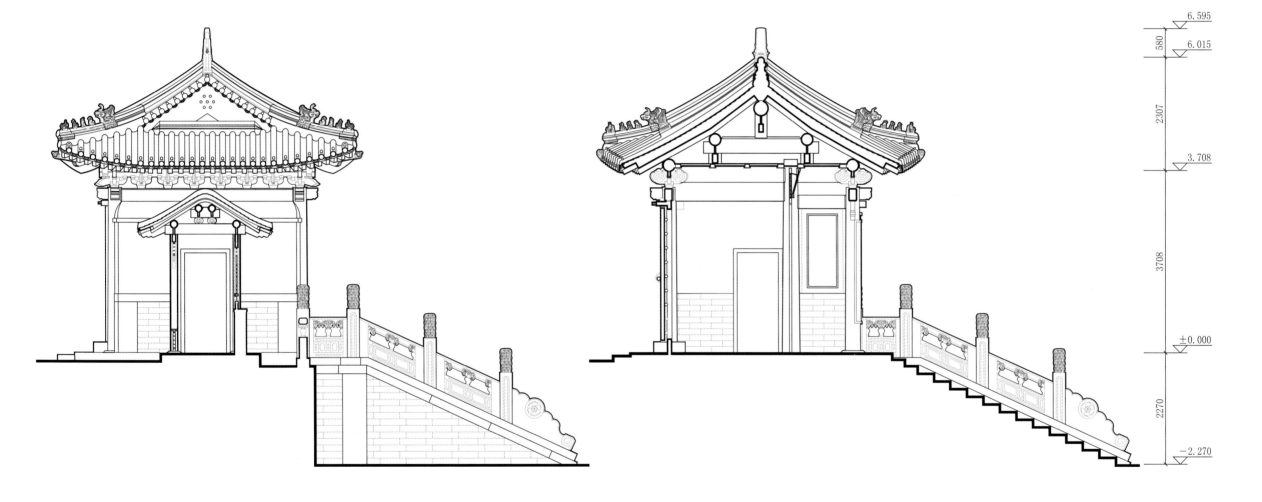

6. 595
580
6. 015
2307
3. 708
3708
±0. 000
2270
-2. 270

850 | 1230 | 1310 | 1330 | 340 | 1055 | 4410
10525

二宫门侧立面图

Side Elevation of the Second Palace Gate

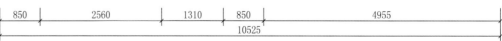

850 | 2560 | 1310 | 850 | 4955
10525

二宫门明间剖面图

Mingjian Section of the Second Palace Gate

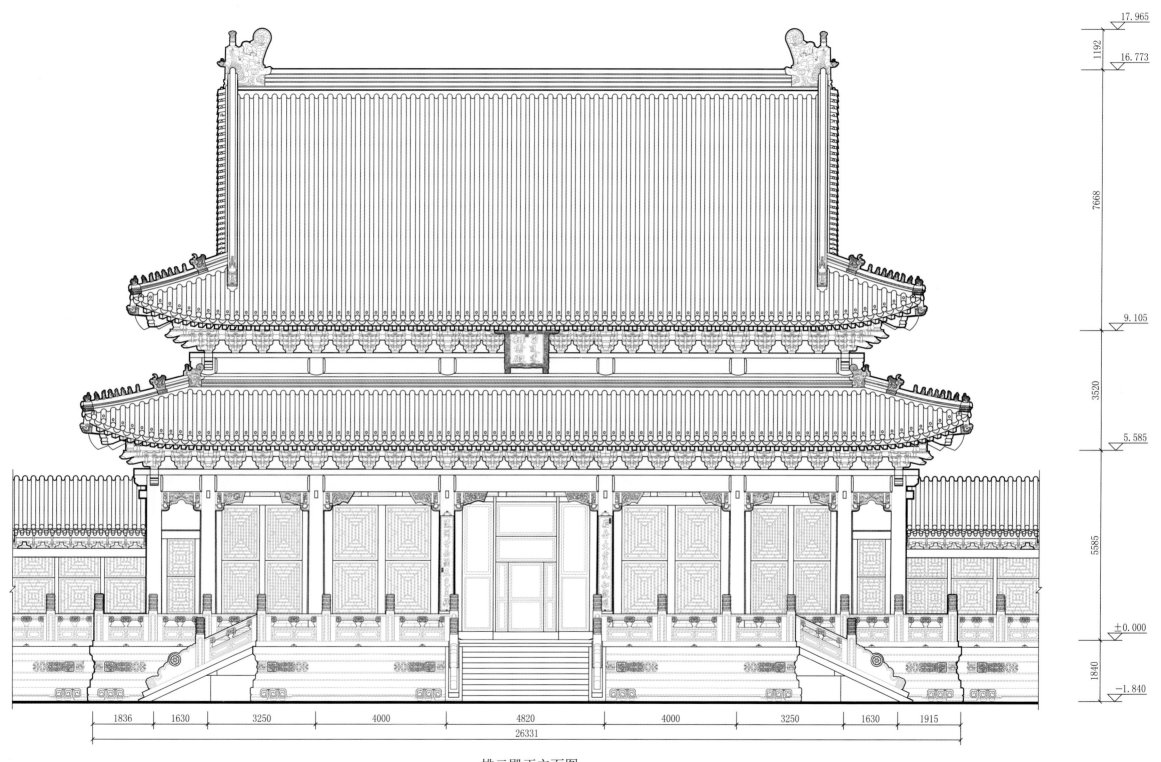

排云殿正立面图

Front Elevation of the *Paiyun* Hall

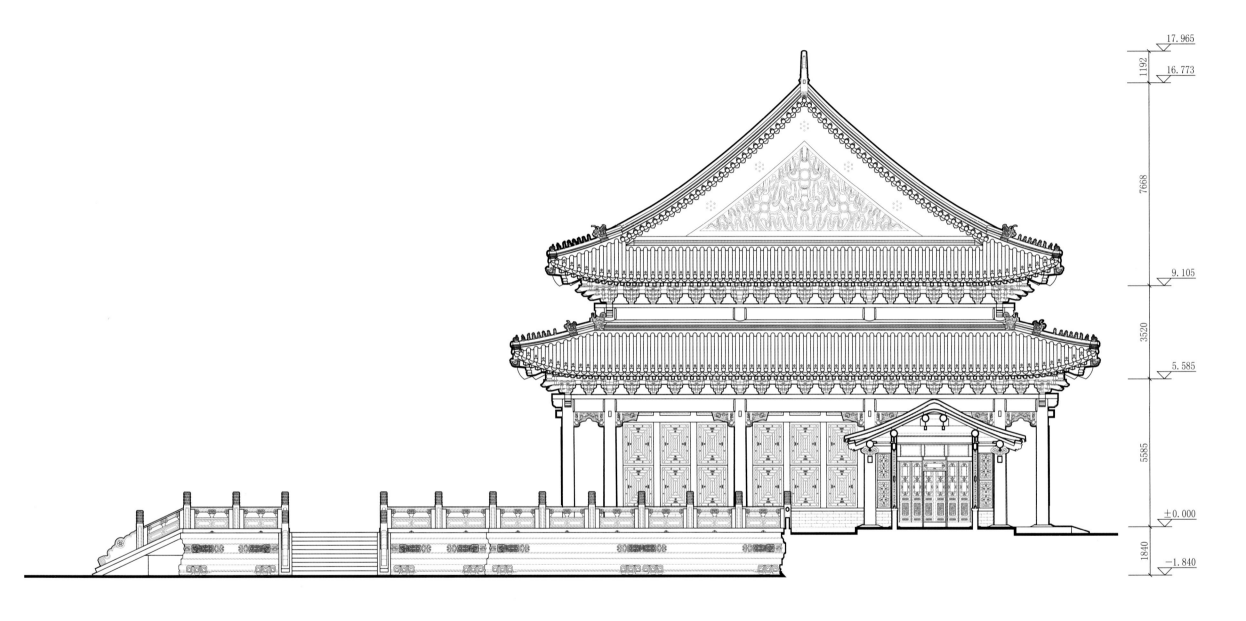

17.965
1192
16.773
7668
9.105
3520
5.585
5585
±0.000
1840
−1.840

3500　　14011　　1127　1630　　7630　　7630　1630　1125
38283

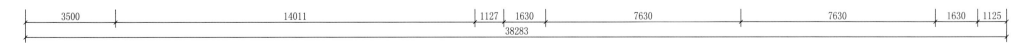

排云殿侧立面图
Side Elevation of the *Paiyun* Hall

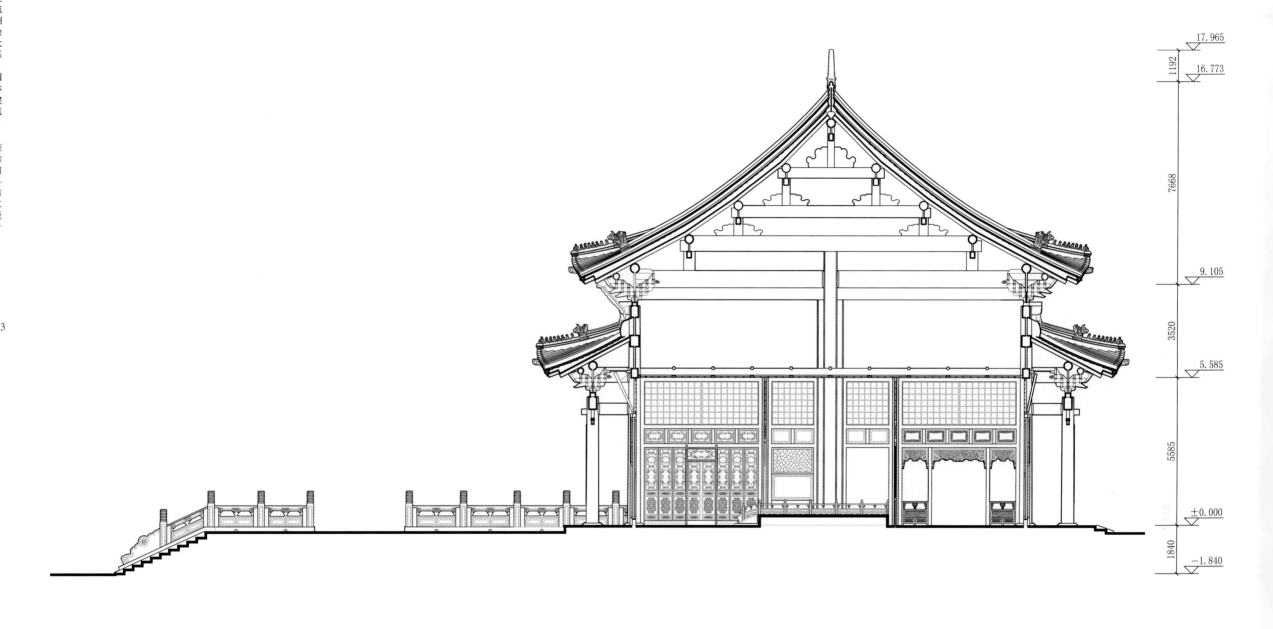

17.965

16.773

1192

7668

9.105

3520

5.585

5585

±0.000

1840

-1.840

3500 14011 1127 1630 7630 7630 1630 1125

38283

排云殿明间剖面图

Mingjian Section of the *Paiyun* Hall

9.427

928

8.499

4464

4.035

4035

±0.000

778

−0.778

893　1300　3200　3490　3800　3490　3200　1300　923

21596

芳辉殿正立面图

Front Elevation of the *Fanghui* Hall

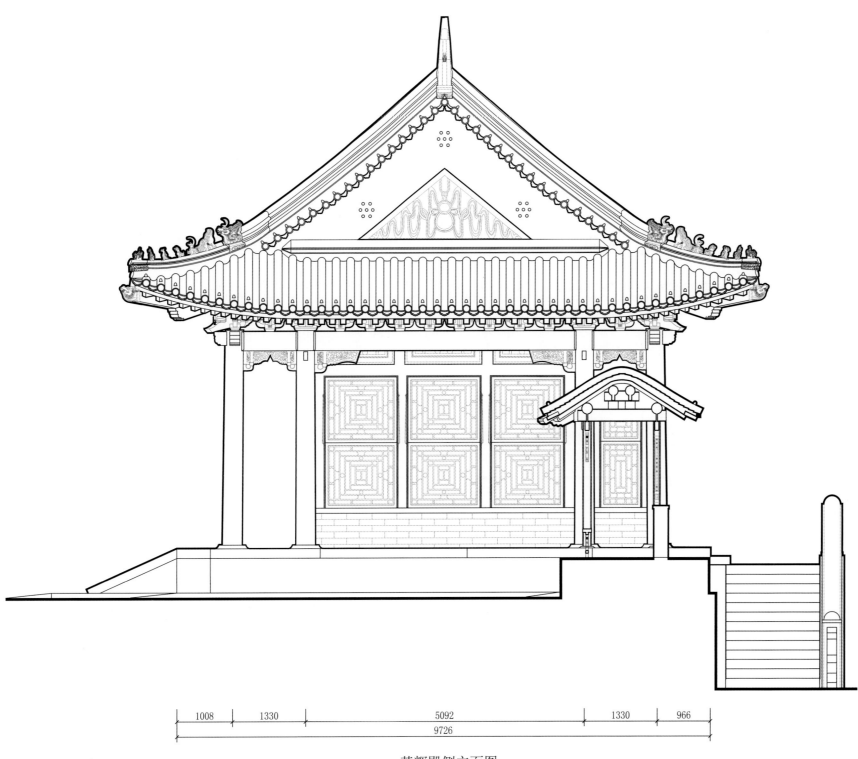

芳辉殿侧立面图
Side Elevation of the *Fanghui* Hall

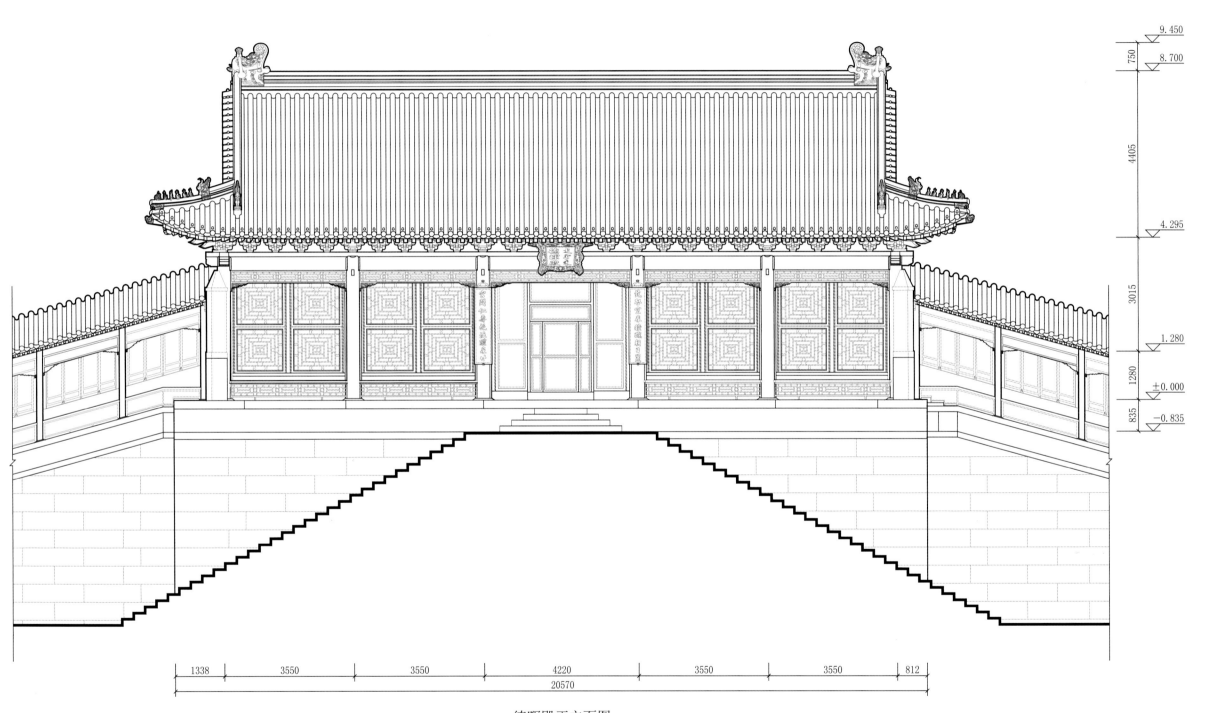

9.450

750

8.700

4405

4.295

3015

1.280

1280

±0.000

835

−0.835

1338　3550　3550　4220　3550　3550　812

20570

德晖殿正立面图

Front Elevation of the *Dehui* Hall

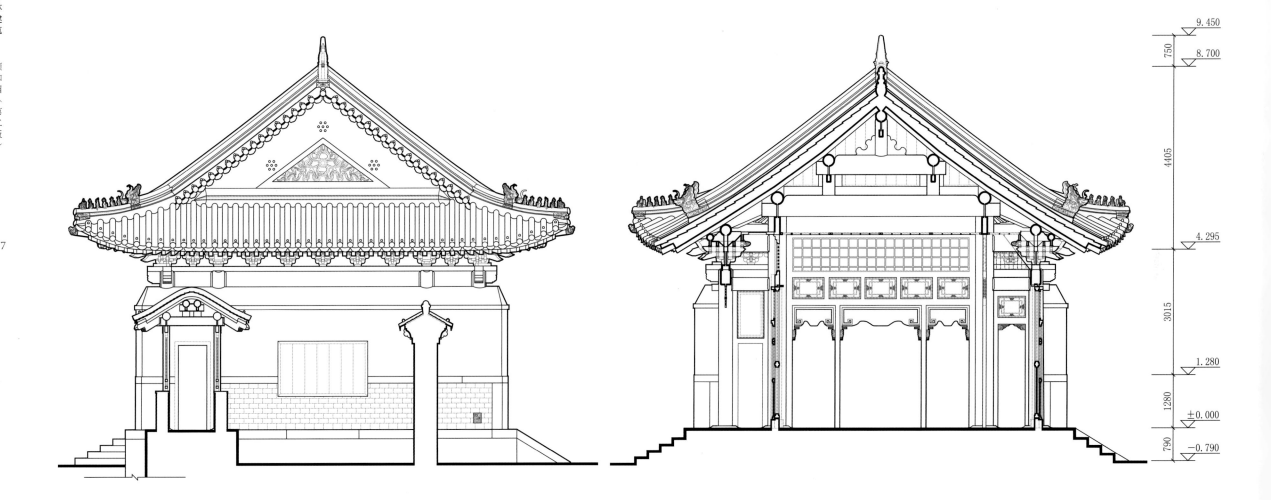

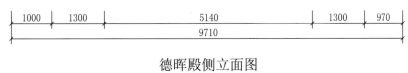

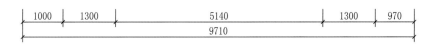

德晖殿侧立面图
Side Elevation of the *Dehui* Hall

德晖殿明间剖面图
Mingjian Section of the *Dehui* Hall

1 南山门 South Gate
2 佛香阁 *Foxiang* Pavilion
3 北山门 North Gate
4 敷华亭 *Fuhua* Pavilion
5 撷秀亭 *Xiexiu* Pavilion
6 众香界牌楼 *Zhongxiangjie* Pailou

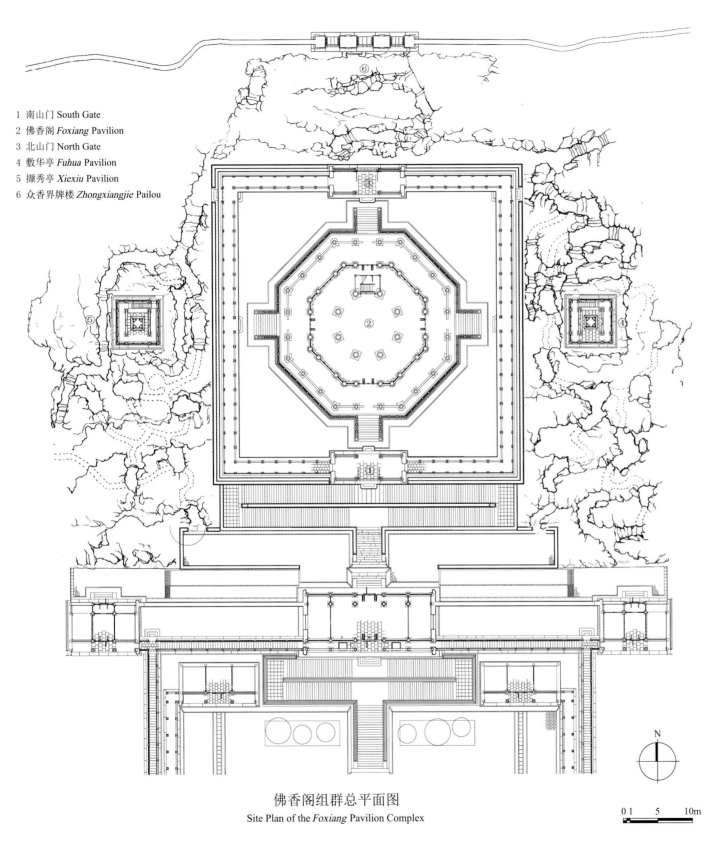

佛香阁组群总平面图
Site Plan of the *Foxiang* Pavilion Complex

0 1   5   10m

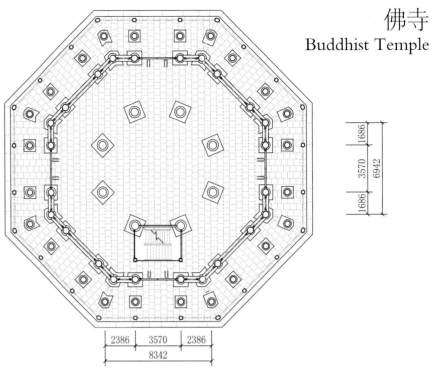

佛香阁二层平面图
Second Floor Plan of the *Foxiang* Pavilion

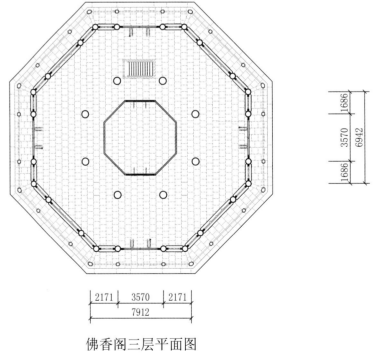

佛香阁三层平面图
Thied Floor Plan of the *Foxiang* Pavilion

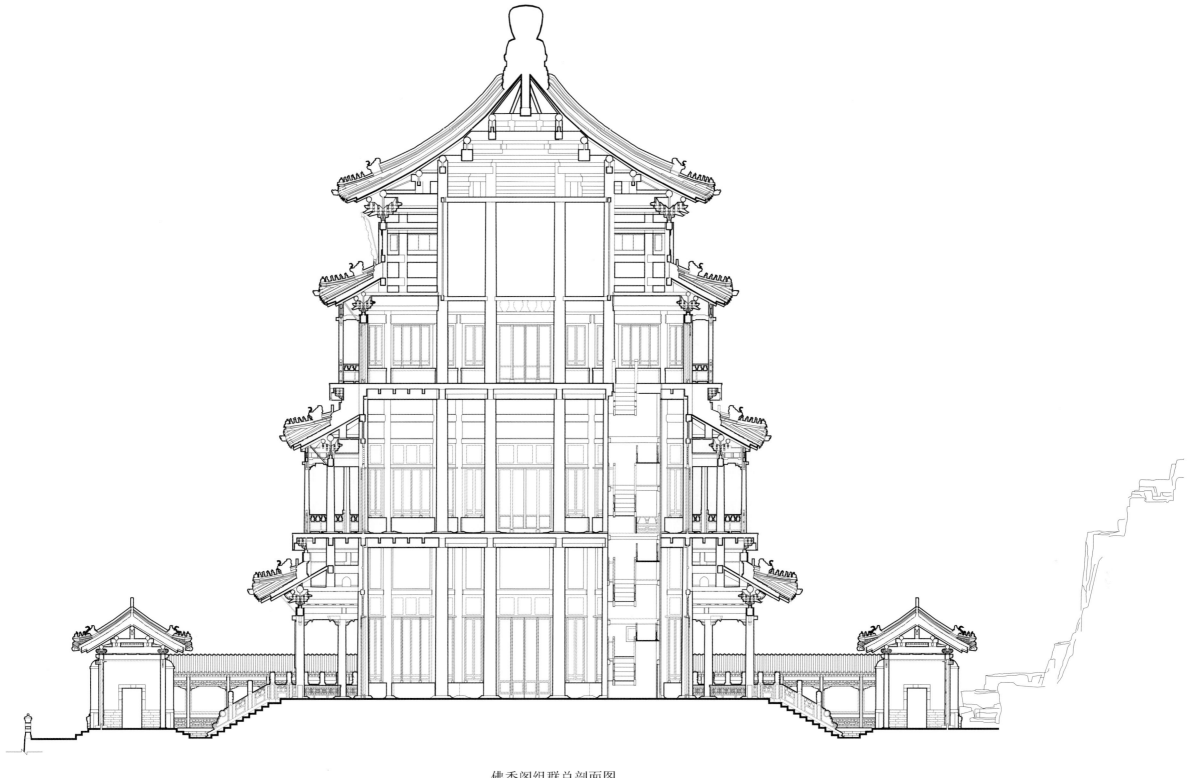

佛香阁组群总剖面图

Section of the *Foxiang* Pavilion Complex

0 1 5m

佛香阁正立面图
Front Elevation of the *Foxiang* Pavilion

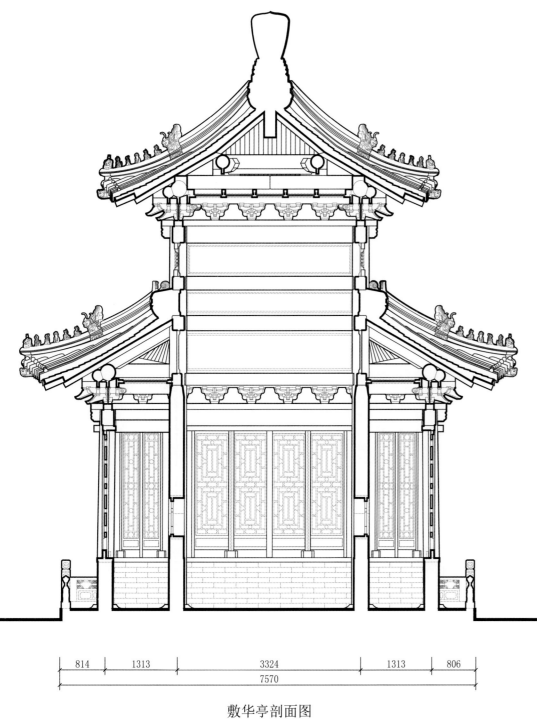

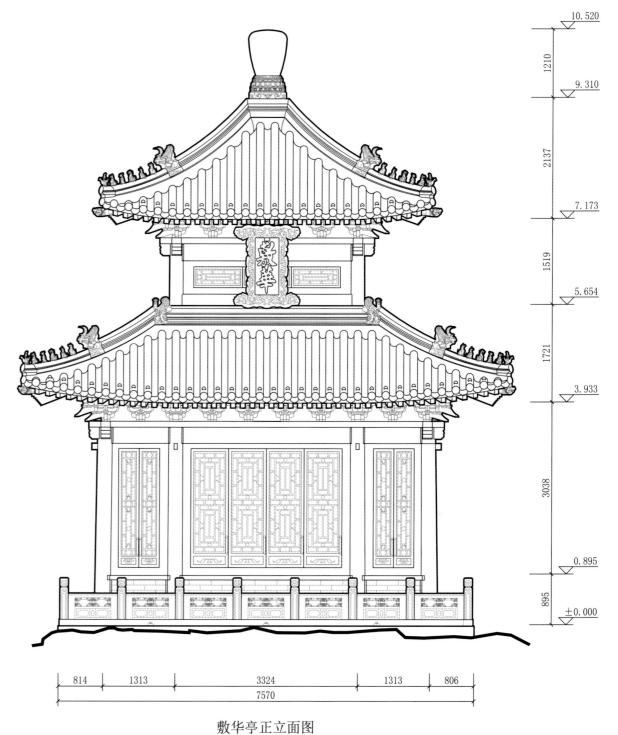

敷华亭剖面图

Section of the *Fuhua* Pavilion

敷华亭正立面图

Front Elevation of the *Fuhua* Pavilion

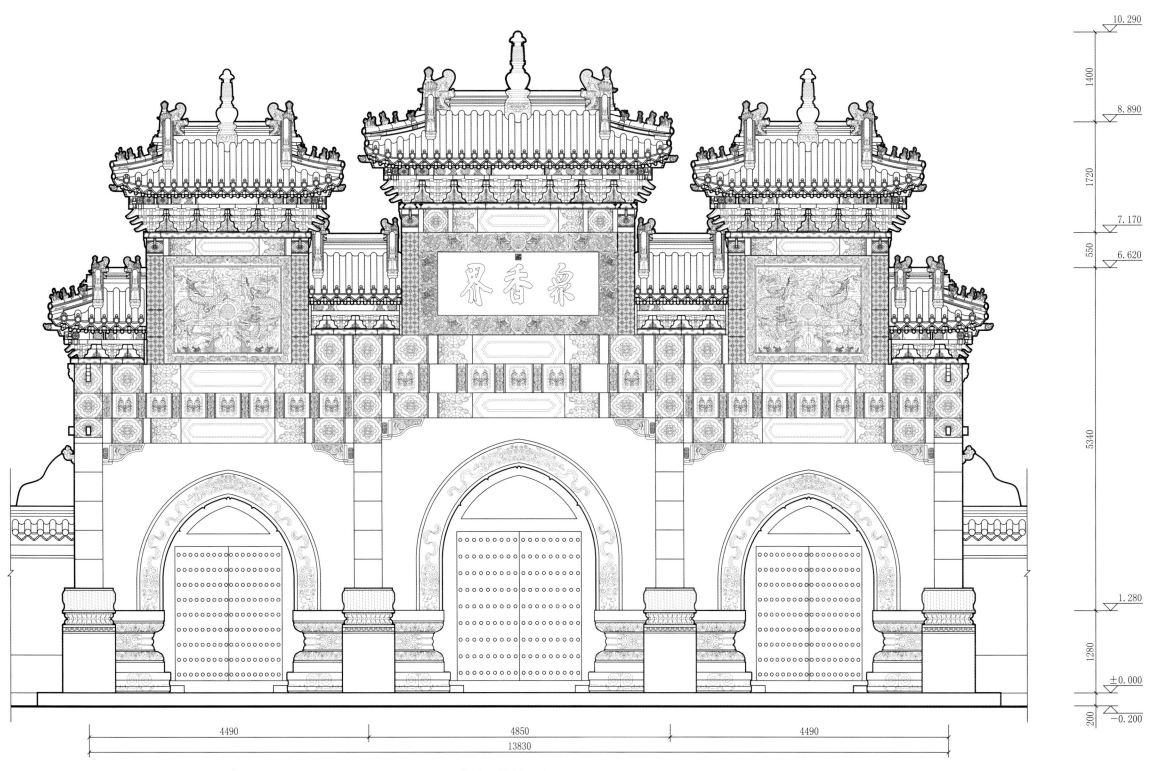

10.290

1400

8.890

1720

7.170

550

6.620

5340

界香众

1.280

1280

±0.000

200 −0.200

4490　　　　　4850　　　　　4490

13830

众香界牌楼正立面图
Front Elevation of the *Zhongxiangjie* Pailou

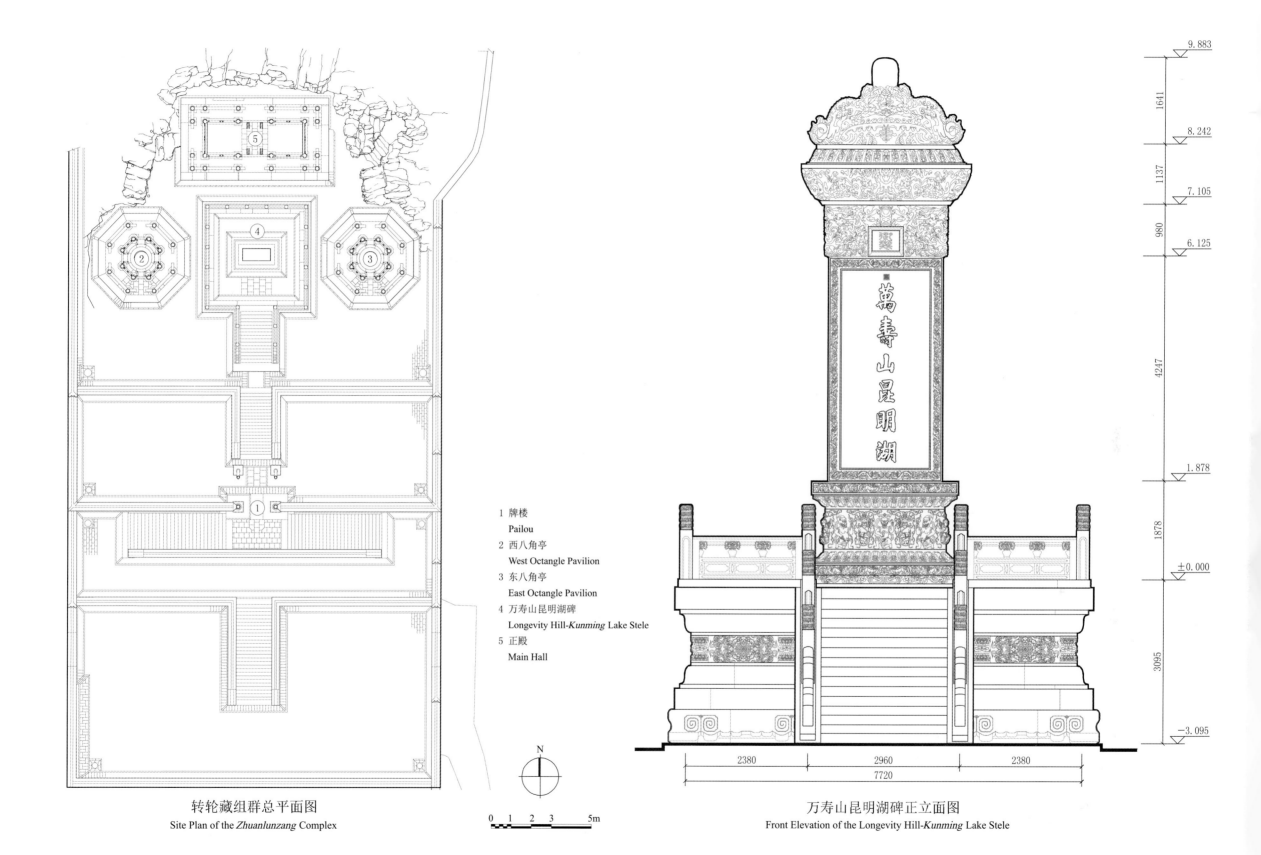

1　牌楼
　　Pailou
2　西八角亭
　　West Octangle Pavilion
3　东八角亭
　　East Octangle Pavilion
4　万寿山昆明湖碑
　　Longevity Hill-*Kunming* Lake Stele
5　正殿
　　Main Hall

N

转轮藏组群总平面图
Site Plan of the *Zhuanlunzang* Complex

0　1　2　3　　5m

9.883
1641
8.242
1137
7.105
980
6.125
4247
1.878
1878
±0.000
3095
−3.095

2380　　2960　　2380
7720

万寿山昆明湖碑正立面图
Front Elevation of the Longevity Hill-*Kunming* Lake Stele

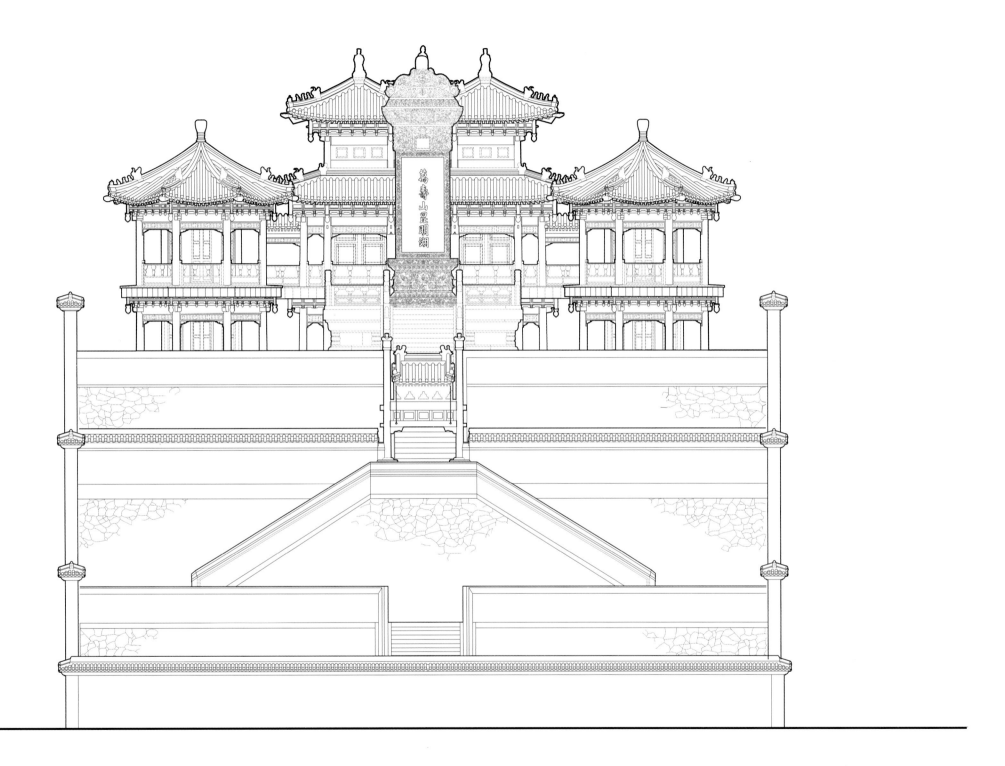

转轮藏组群总立面图

Elevation of the *Zhuanlunzang* Complex

0　1　　　　　5m

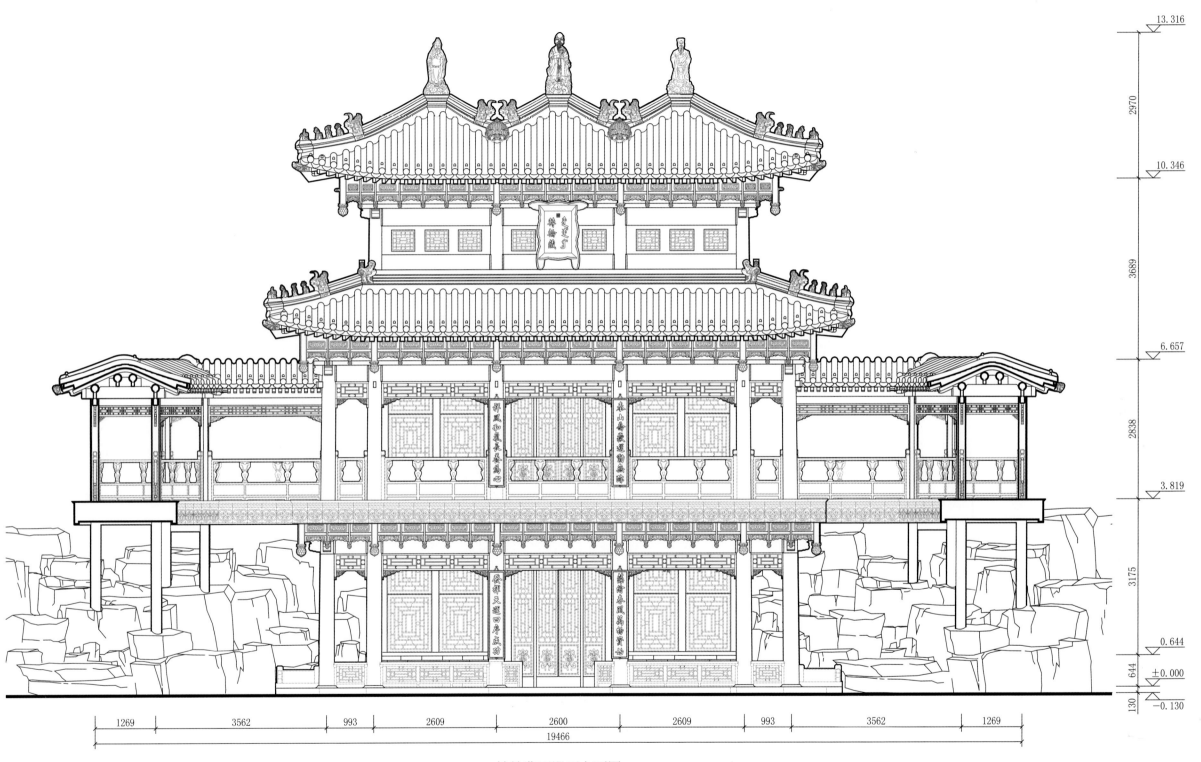

转轮藏正殿正立面图

Front Elevation of the Main Hall of *Zhuanlunzang*

转轮藏正殿侧立面图
Side Elevation of the Main Hall of *Zhuanlunzang*

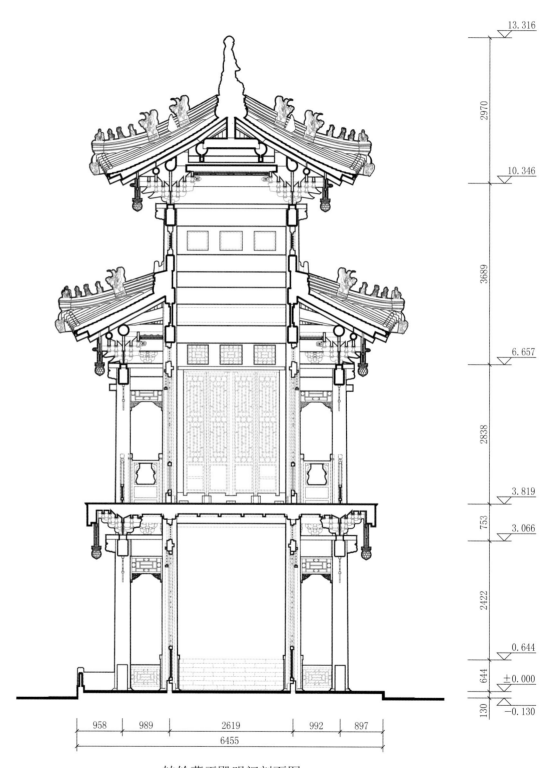

转轮藏正殿明间剖面图
Mingjian Section of the Main Hall of *Zhuanlunzang*

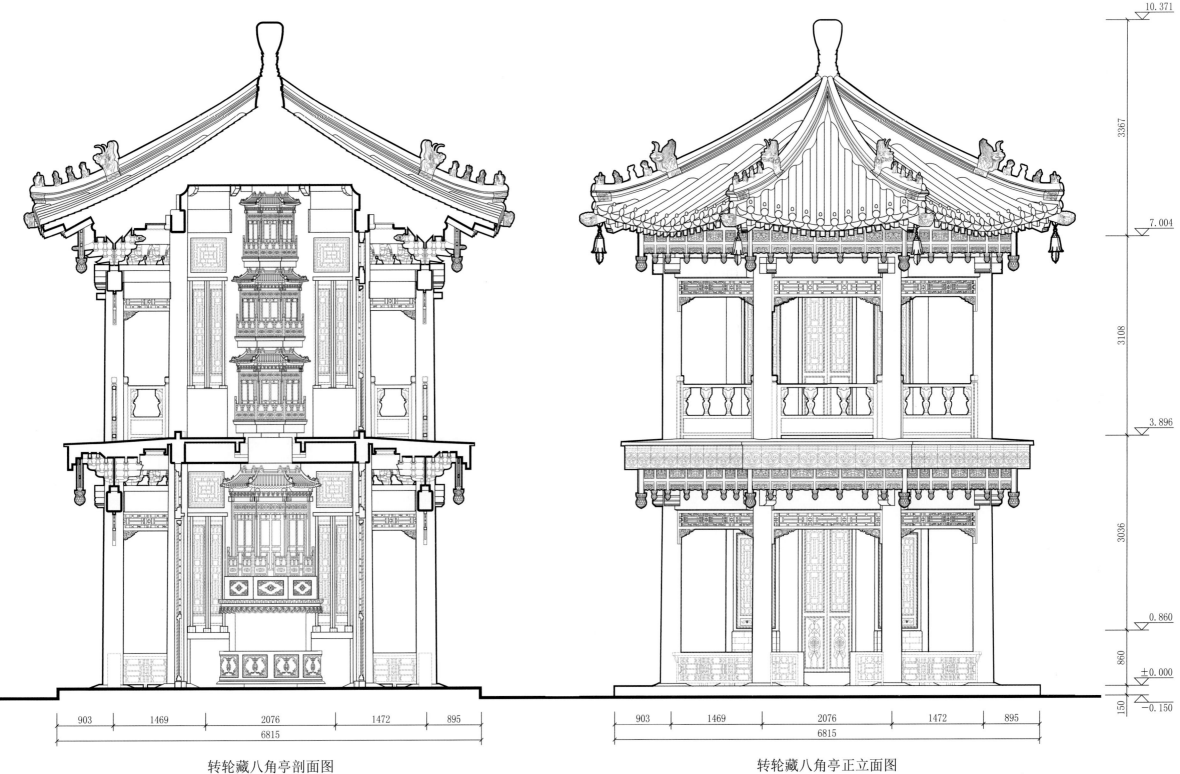

转轮藏八角亭剖面图
Section of the Octangle Pavilion of *Zhuanlunzang*

转轮藏八角亭正立面图
Front Elevation of the Octangle Pavilion of *Zhuanlunzang*

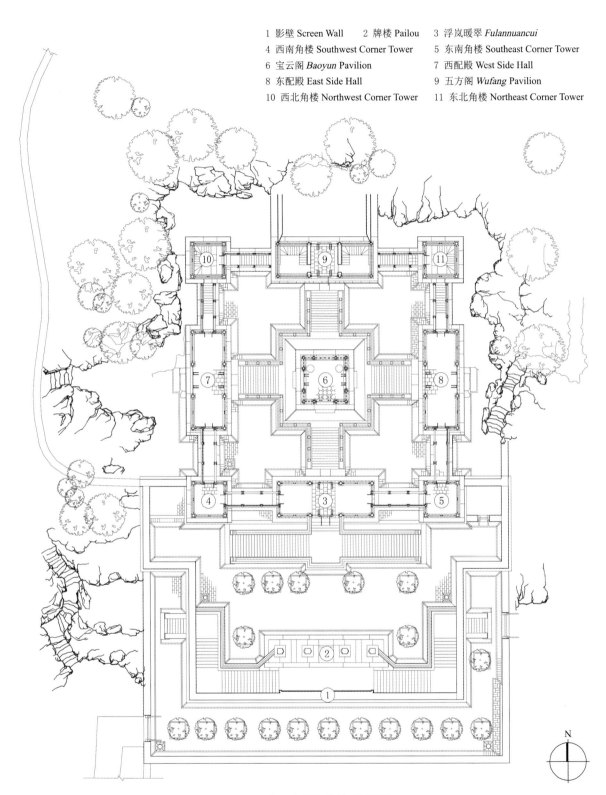

宝云阁组群总平面图
Site Plan of the *Baoyun* Pavilion Complex

0 1 2 3　5m

N

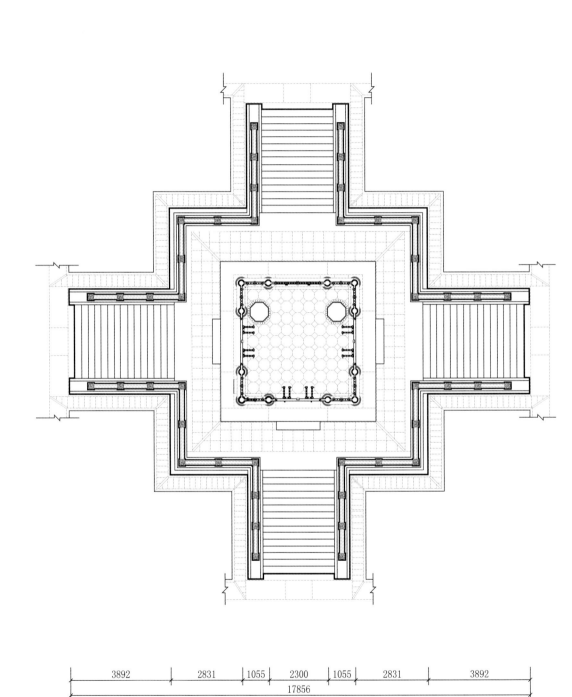

| 3892 | 2831 | 1055 | 2300 | 1055 | 2831 | 3892 |

17856

宝云阁铜亭平面图
Plan of the *Baoyun* Copper Pavilion

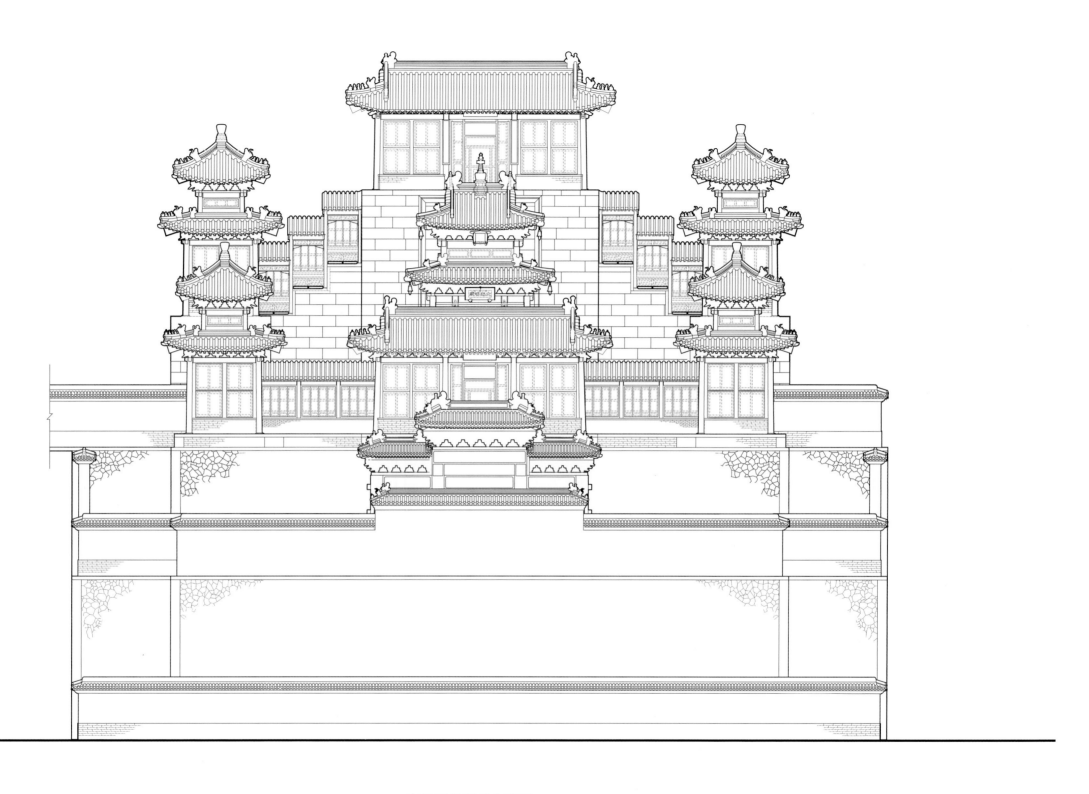

宝云阁组群总立面图

Elevation of the *Baoyun* Pavilion Complex

8.325

446

7.879

1185

6.694

482

6.212

859

5.353

3732

1.621

1621

±0.000

暮霭朝岚常自写

无边清况愦幽襟　泉声入目凉

物舍妙理揔堪寻

境自远尘皆入梦　山色园心远

我许崇情记远跡

1620　2890　3880　2890　1620

12900

宝云阁牌楼正立面图

Front Elevation of the Pailou of *Baoyun* Pavilion

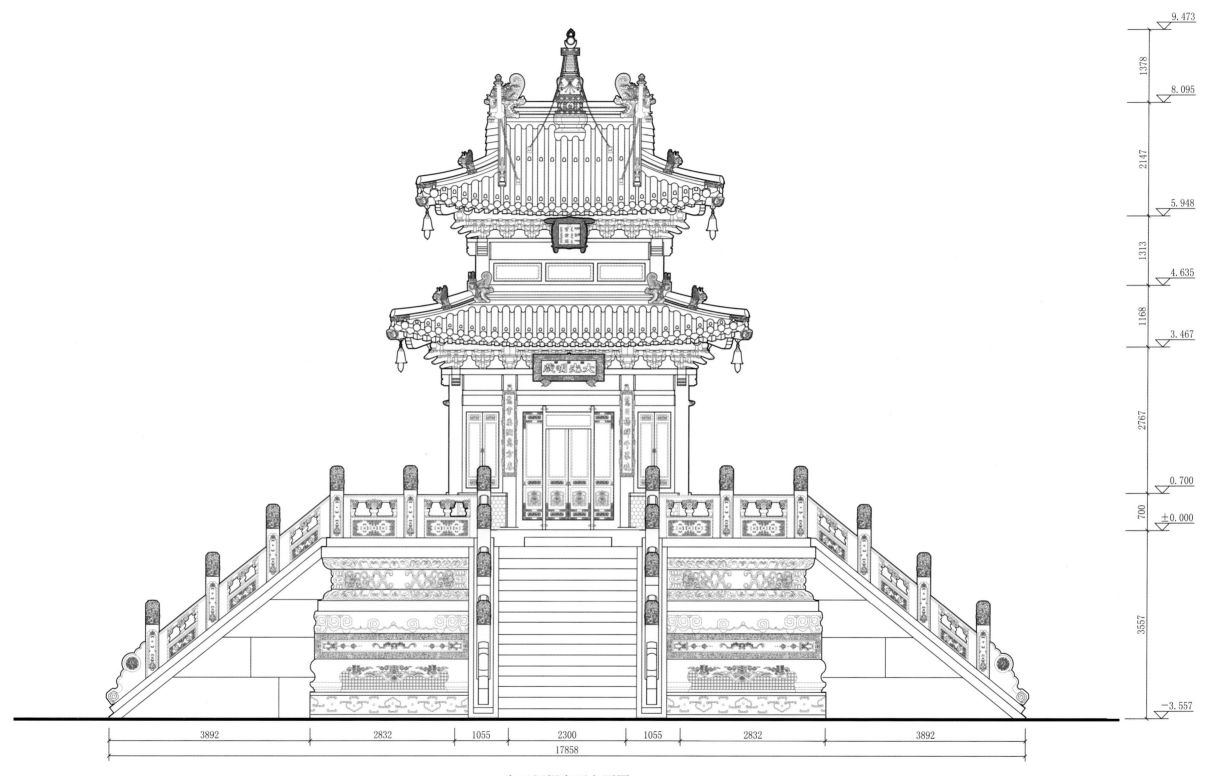

9.473

1378

8.095

2147

5.948

1313

4.635

1168

3.467

2767

0.700

700

±0.000

3557

−3.557

3892 2832 1055 2300 1055 2832 3892

17858

宝云阁铜亭正立面图

Front Elevation of the *Baoyun* Copper Pavilion

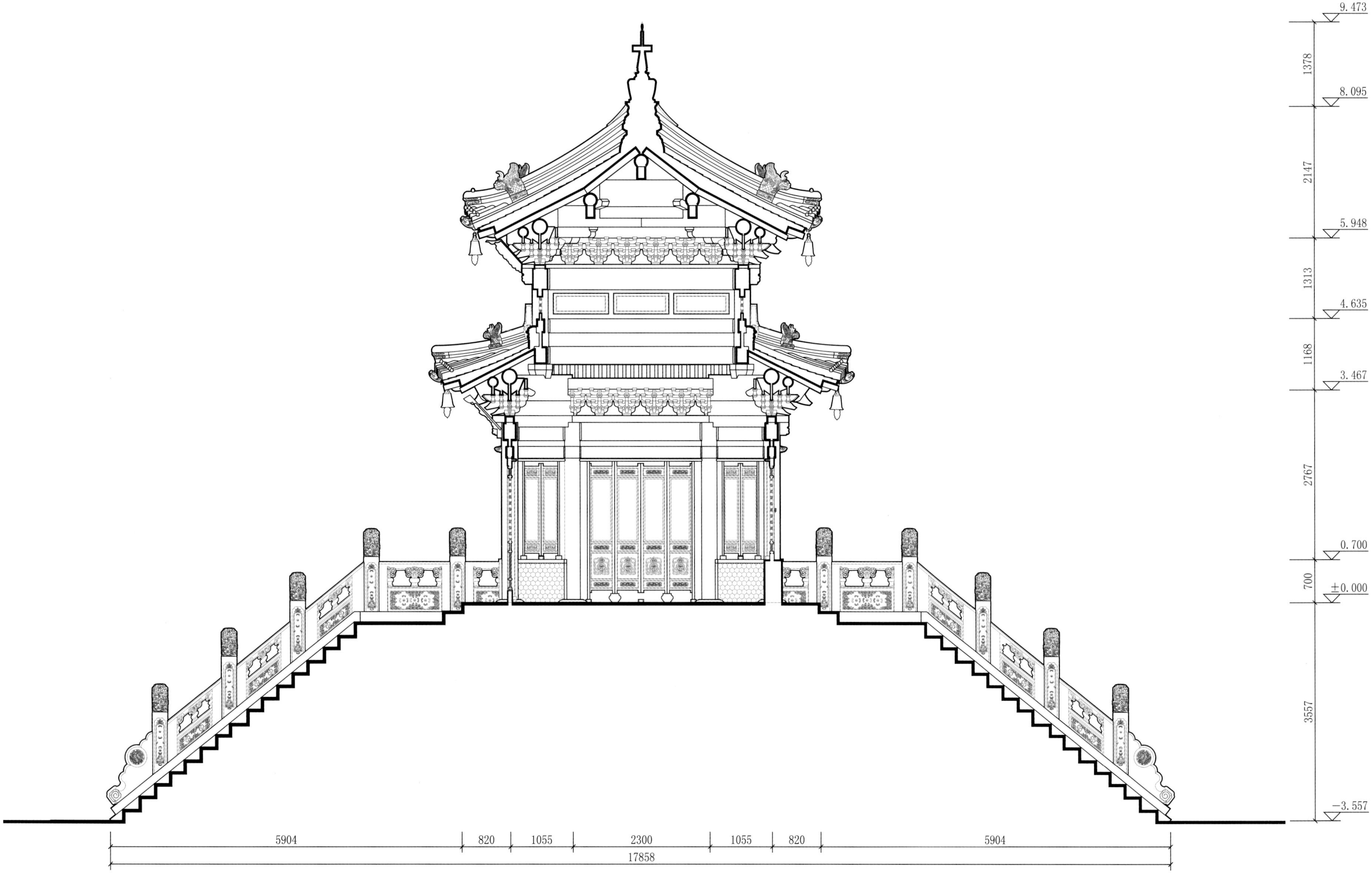

9.473

1378

8.095

2147

5.948

1313

4.635

1168

3.467

2767

0.700

700 ±0.000

3557

−3.557

5904 | 820 | 1055 | 2300 | 1055 | 820 | 5904

17858

宝云阁铜亭剖面图

Section of the *Baoyun* Copper Pavilion

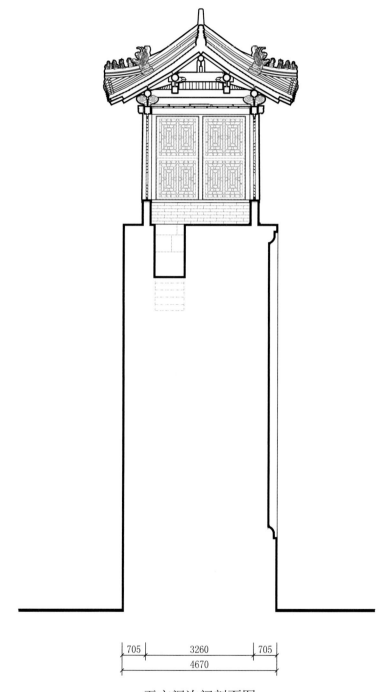

705 | 3260 | 705
4670

五方阁次间剖面图
Cijian Section of the *Wufang* Pavilion

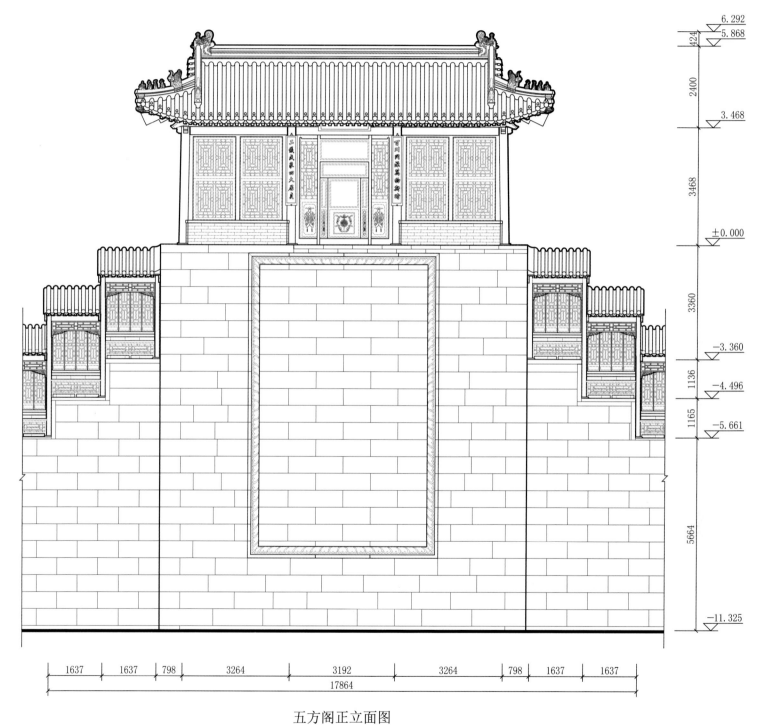

6.292
5.868
424
2400
3.468
3468
±0.000
3360
−3.360
1136
−4.496
1165
−5.661
5664
−11.325

1637 | 1637 | 798 | 3264 | 3192 | 3264 | 798 | 1637 | 1637
17864

五方阁正立面图
Front Elevation of the *Wufang* Pavilion

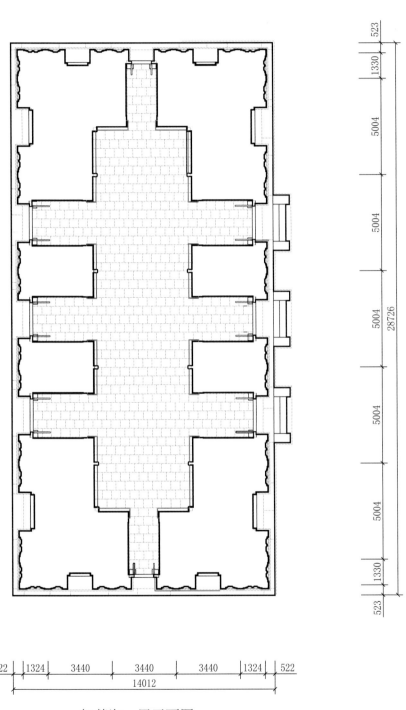

智慧海一层平面图
First Floor Plan of the *Zhihuihai*

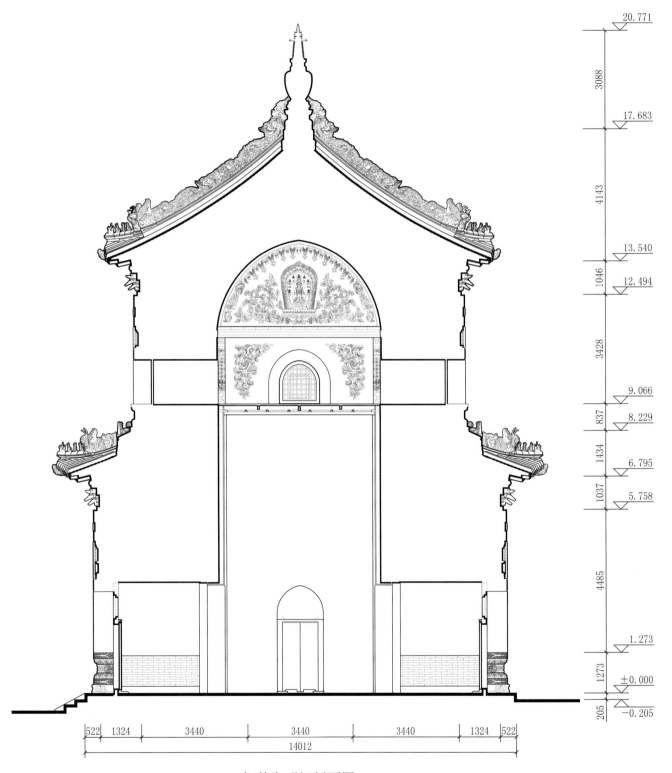

智慧海明间剖面图
Mingjian Section of the *Zhihuihai*

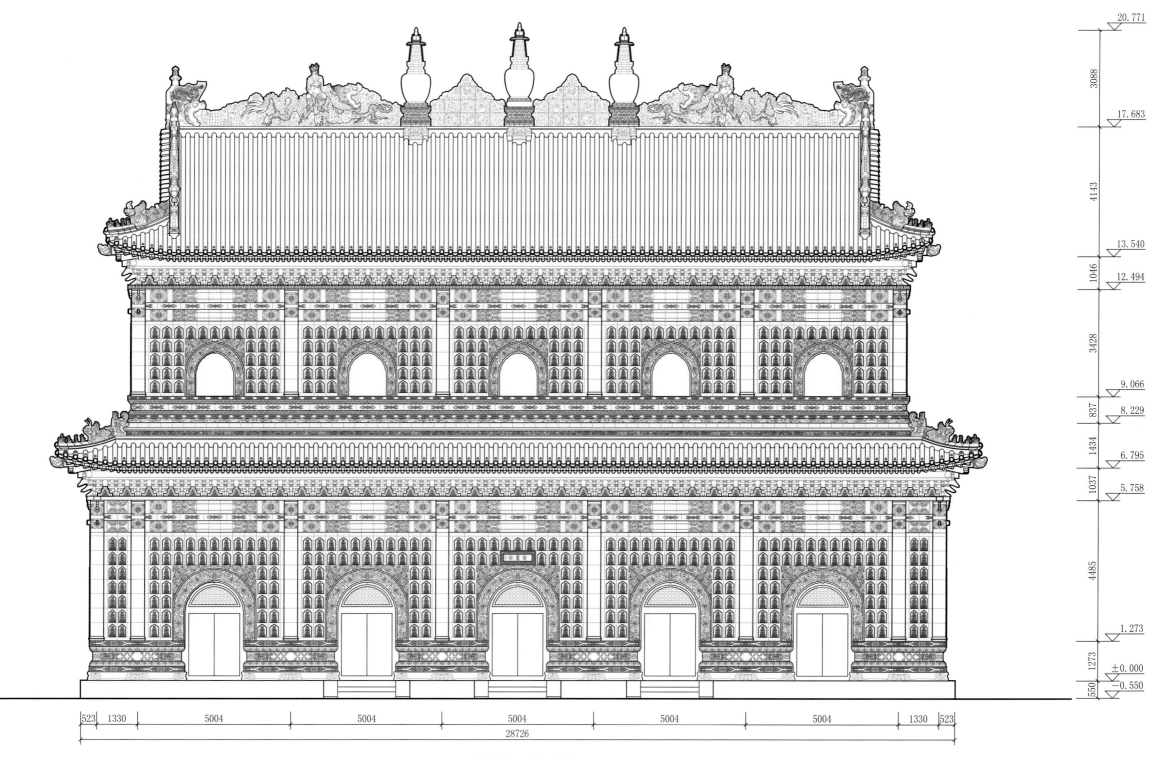

20.771

3088

17.683

4143

13.540

1046

12.494

3428

9.066

837

8.229

1434

6.795

1037

5.758

4485

1.273

1273

±0.000

550

−0.550

523 1330 5004 5004 5004 5004 5004 1330 523

28726

智慧海正立面图

Front Elevation of the *Zhihuihai*

智慧海二层山墙及月光墙大样图

Gable and Moon Wall of the Second Floor of *Zhihuihai*

0   0.2   1m

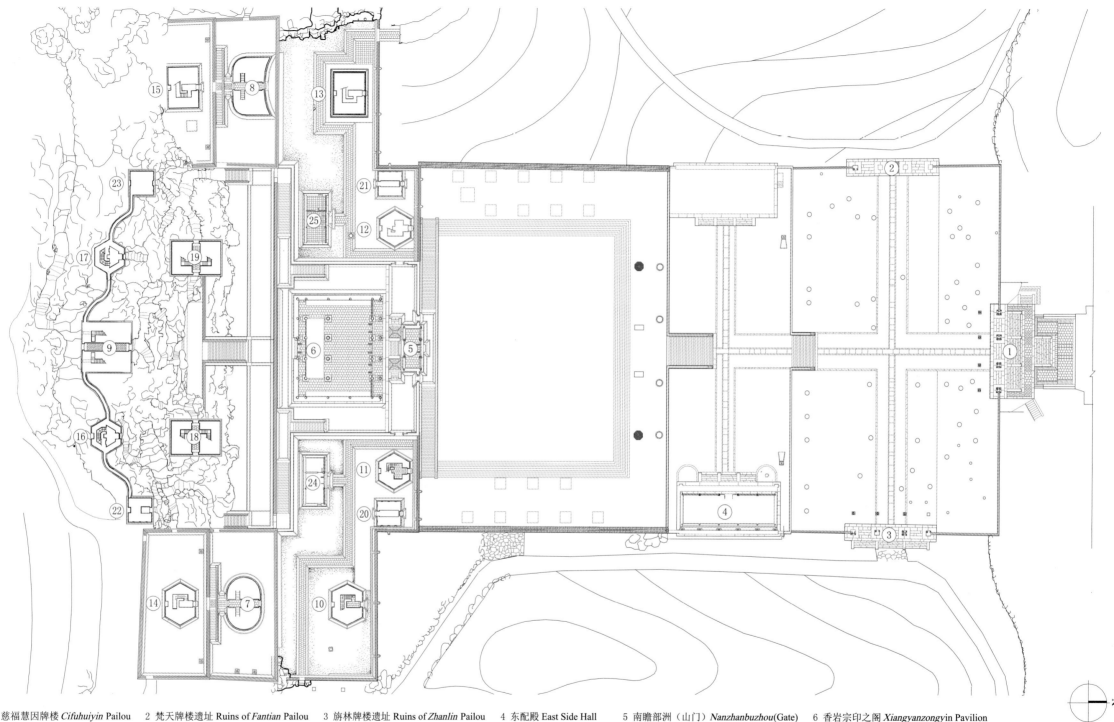

1 慈福慧因牌楼 *Cifuhuiyin* Pailou   2 梵天牌楼遗址 Ruins of *Fantian* Pailou   3 旃林牌楼遗址 Ruins of *Zhanlin* Pailou   4 东配殿 East Side Hall   5 南瞻部洲（山门）*Nanzhanbuzhou*(Gate)   6 香岩宗印之阁 *Xiangyanzongyin* Pavilion

7 西牛贺洲 *Xi'niuhezhou*   8 东胜神洲 *Dongshengshenzhou*   9 北俱芦洲 *Beijuluzhou*   10 一小部洲 *Yixiaobuzhou*   11 二小部洲 *Erxiaobuzhou*   12 三小部洲 *Sanxiaobuzhou*

13 四小部洲 *Sixiaobuzhou*   14 五小部洲 *Wuxiaobuzhou*   15 六小部洲 *Liuxiaobuzhou*   16 七小部洲 *Qixiaobuzhou*   17 八小部洲 *Baxiaobuzhou*   18 日殿 Sun Hall

19 月殿 Moon Hall   20 绿塔 Green Tower   21 红塔 Red Tower   22 白塔 White Tower   23 黑塔 Black Tower

24 东值房 East Duty Room   25 西值房 west Duty Room

四大部洲组群总平面图

Site Plan of the *Sidabuzhou* Complex

0 1　　　5　　　　10m

四大部洲组群总立面图

Front Elevation of the *Sidabuzhou* Complex

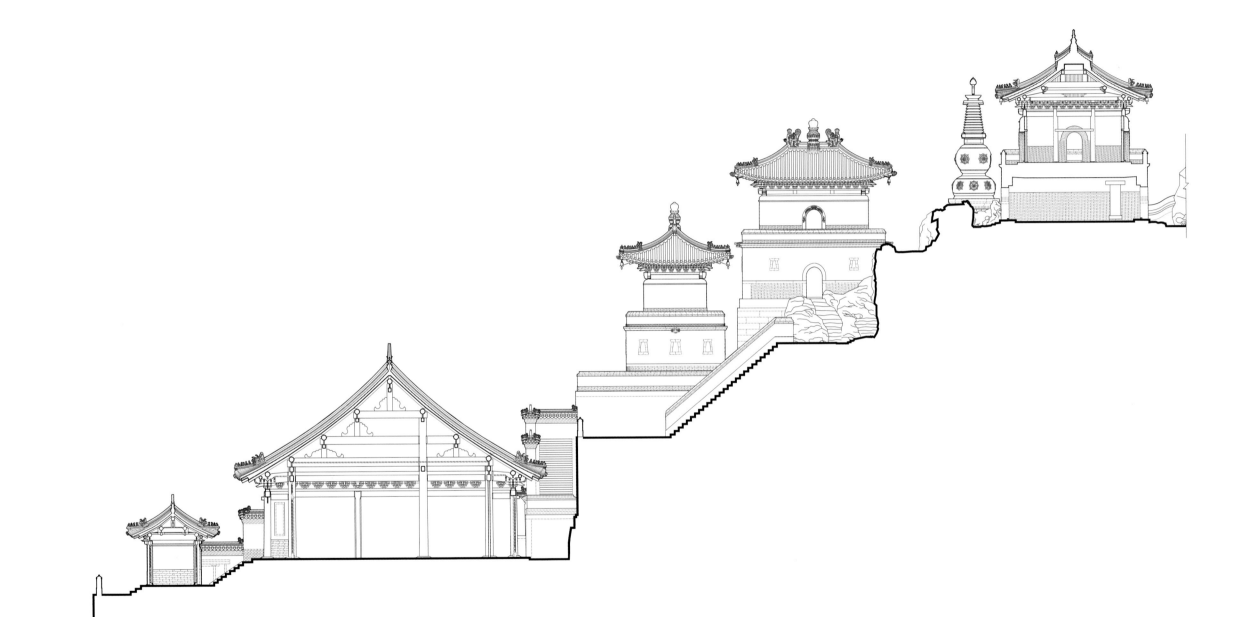

四大部洲组群总剖面图

Section of the *Sidabuzhou* Complex

0 1   5   10m

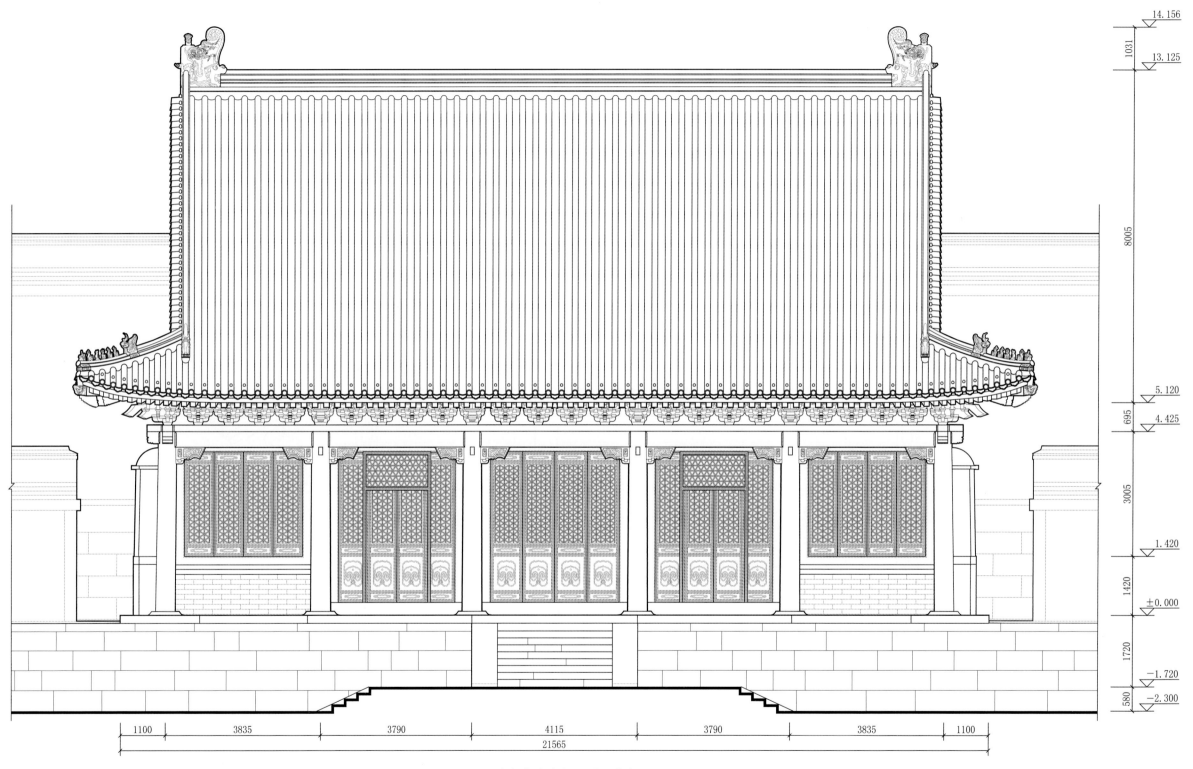

14.156

1031

13.125

8005

5.120

695

4.425

3005

1.420

1420

±0.000

1720

-1.720

580

-2.300

1100   3835   3790   4115   3790   3835   1100

21565

香岩宗印之阁正立面图

Front Elevation of the *Xiangyanzongyin* Pavilion

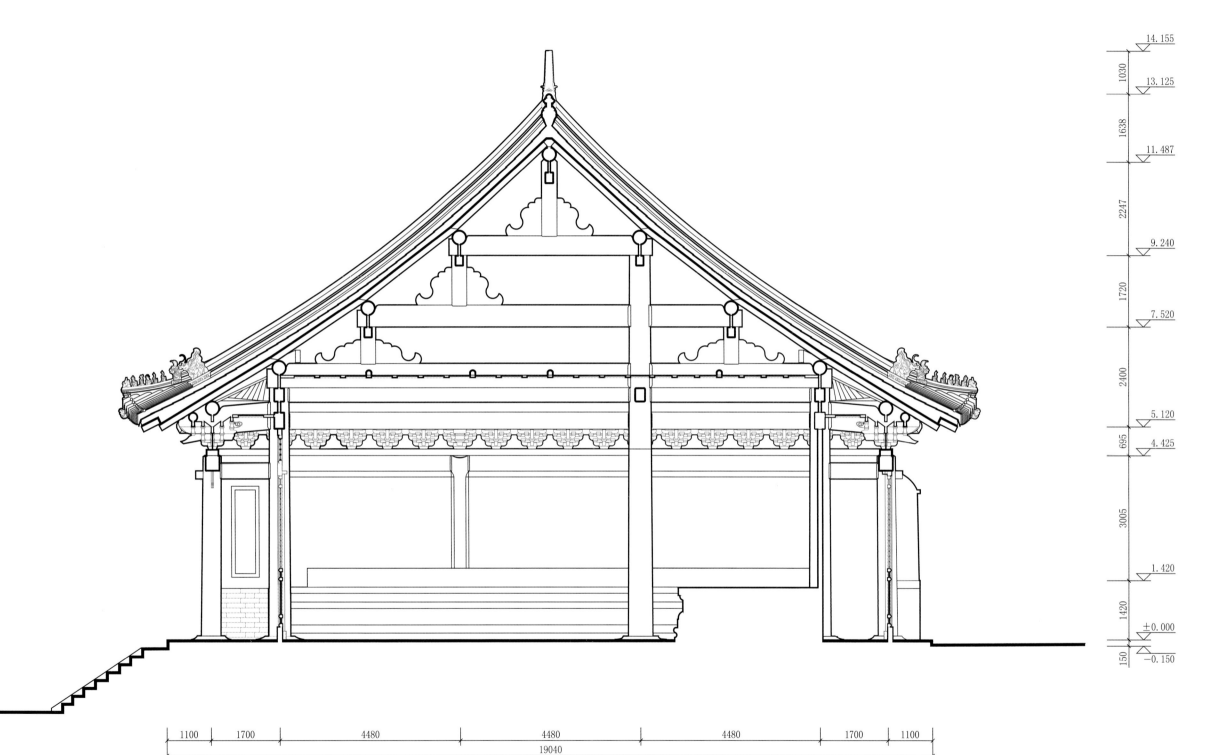

14. 155

1030

13. 125

1638

11. 487

2247

9. 240

1720

7. 520

2400

5. 120

695

4. 425

3005

1. 420

1420

±0. 000

150

−0. 150

1100　1700　4480　4480　4480　1700　1100

19040

香岩宗印之阁明间剖面图

Mingjian Section of the *Xiangyanzongyin* Pavilion

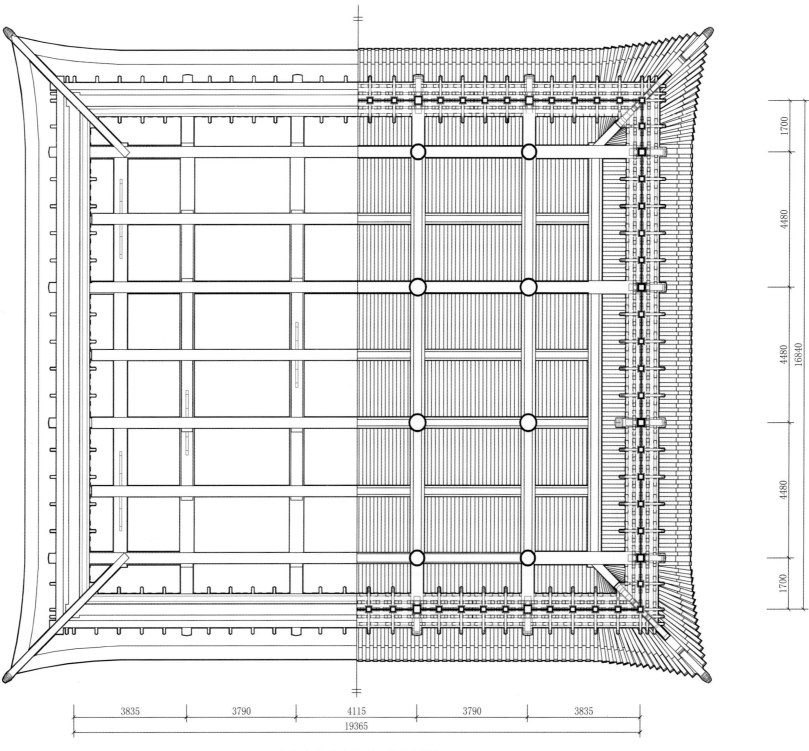

香岩宗印之阁梁架俯仰视图
Framework Plan of the *Xiangyanzongyin* Pavilion

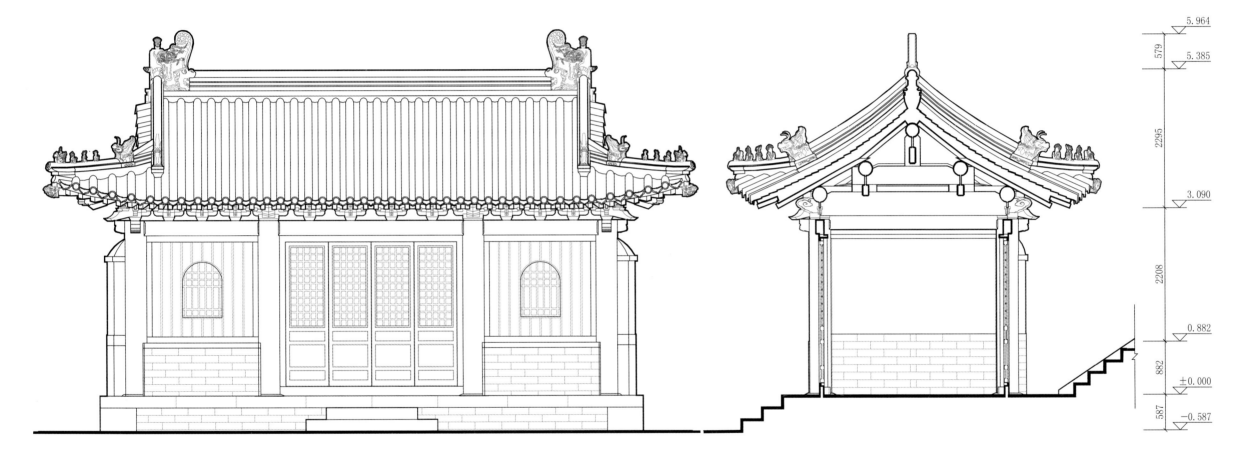

南瞻部洲（山门）侧立面图
Side Elevation of the *Nanzhanbuzhou*(Gate)

南瞻部洲（山门）明间剖面图
Mingjian Section of the *Nanzhanbuzhou*(Gate)

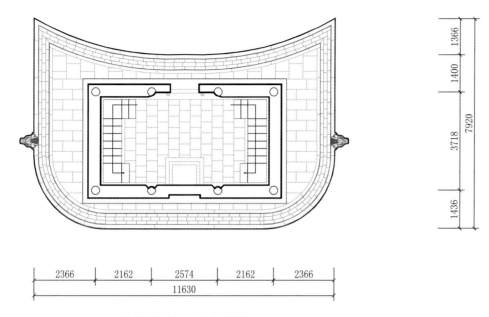

东胜神洲二层平面图

Second Floor Plan of the *Dongshengshenzhou*

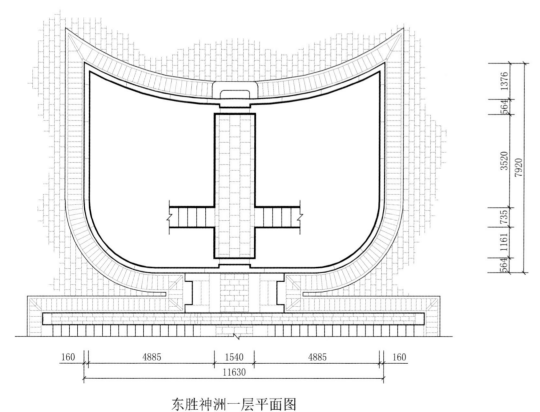

东胜神洲一层平面图

First Floor Plan of the *Dongshengshenzhou*

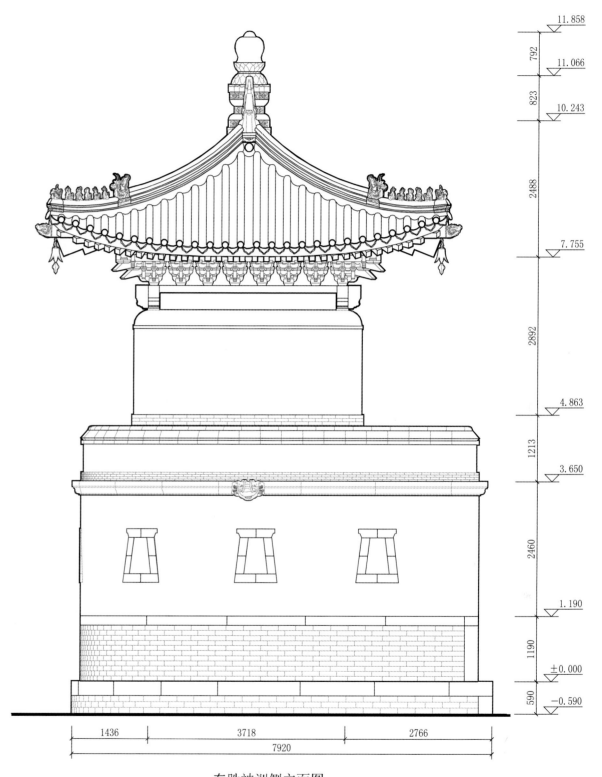

东胜神洲侧立面图

Side Elevation of the *Dongshengshenzhou*

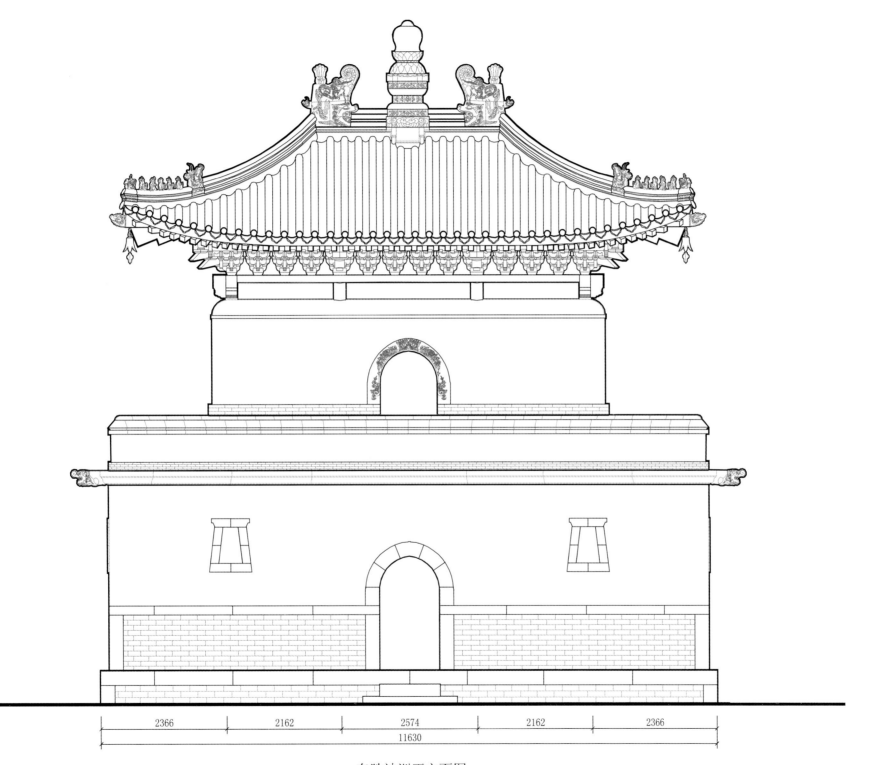

11.858

792

11.066

823

10.243

2488

7.755

2892

4.863

1213

3.650

2460

1.190

1190

±0.000

590

−0.590

2366　2162　2574　2162　2366

11630

东胜神洲正立面图

Front Elevation of the *Dongshengshenzhou*

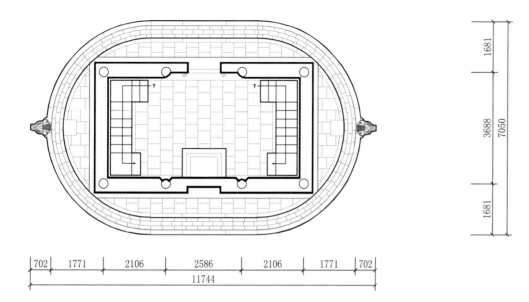

西牛贺洲二层平面图

Second Floor Plan of the *Xi'niuhezhou*

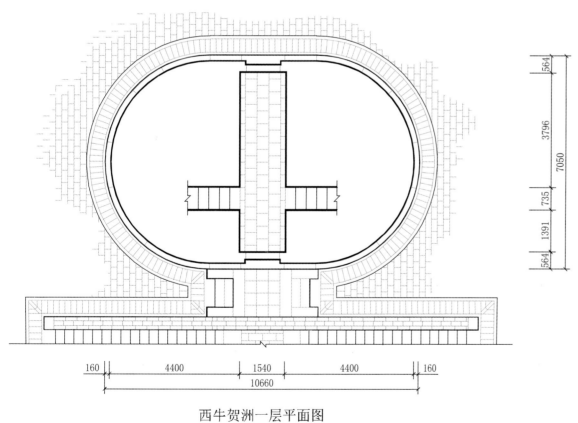

西牛贺洲一层平面图

First Floor Plan of the *Xi'niuhezhou*

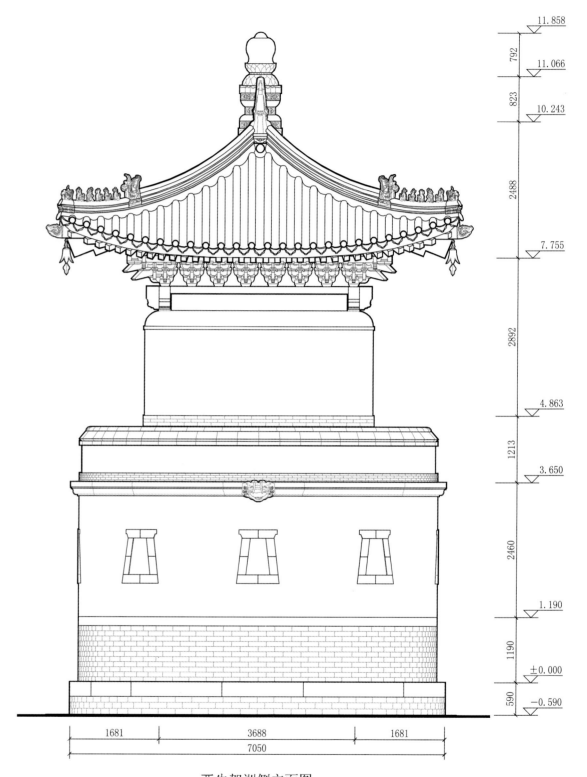

西牛贺洲侧立面图

Side Elevation of the *Xi'niuhezhou*

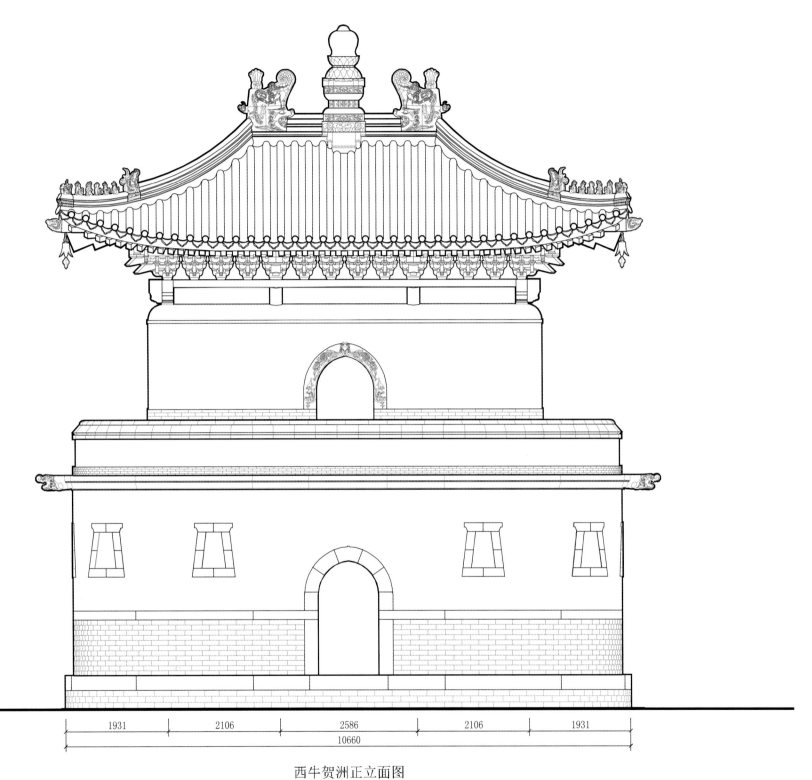

11.858

792

11.066

823

10.243

2488

7.755

2892

4.863

1213

3.650

2460

1.190

1190

±0.000

590

−0.590

1931　2106　2586　2106　1931

10660

西牛贺洲正立面图

Front Elevation of the *Xi'niuhezhou*

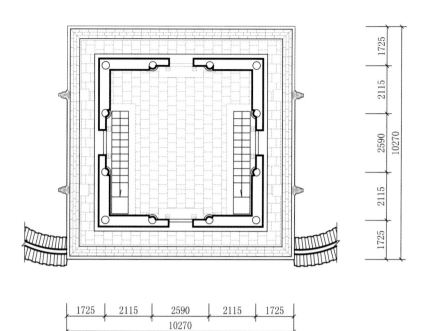

北俱芦洲二层平面图

Second Floor Plan of the *Beijuluzhou*

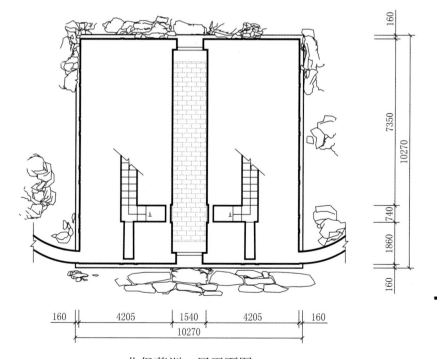

北俱芦洲一层平面图

First Floor Plan of the *Beijuluzhou*

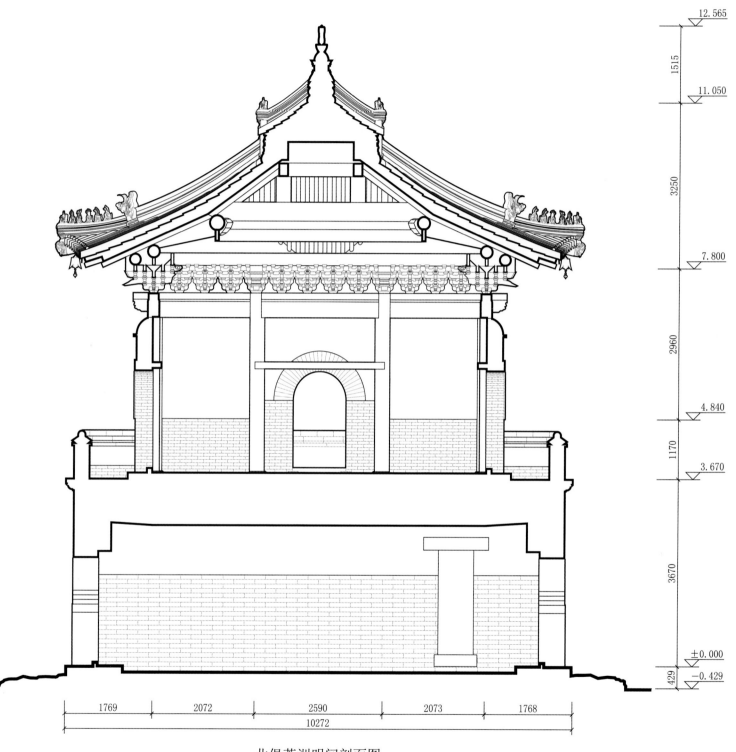

北俱芦洲明间剖面图

Mingjian Section of the *Beijuluzhou*

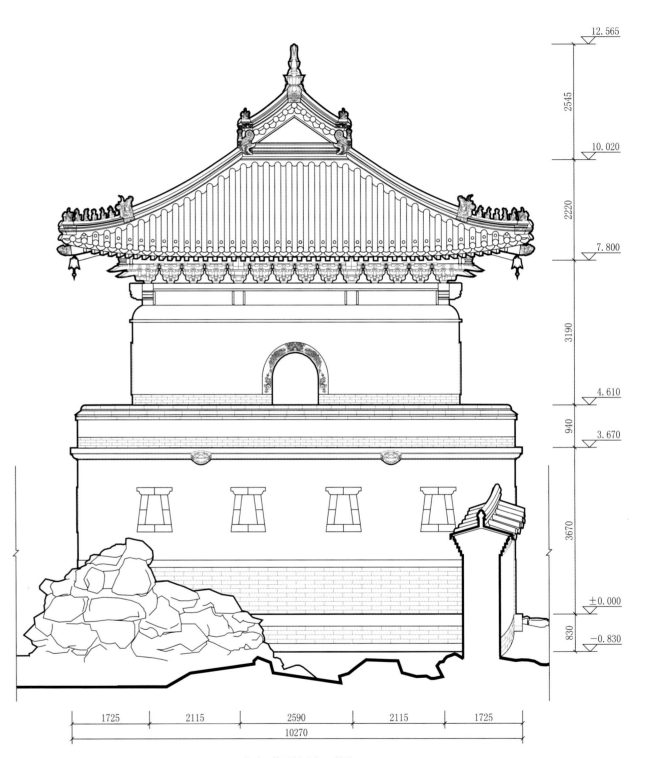

北俱芦洲正立面图
Front Elevation of the *Beijuluzhou*

北俱芦洲侧立面图
Side Elevation of the *Beijuluzhou*

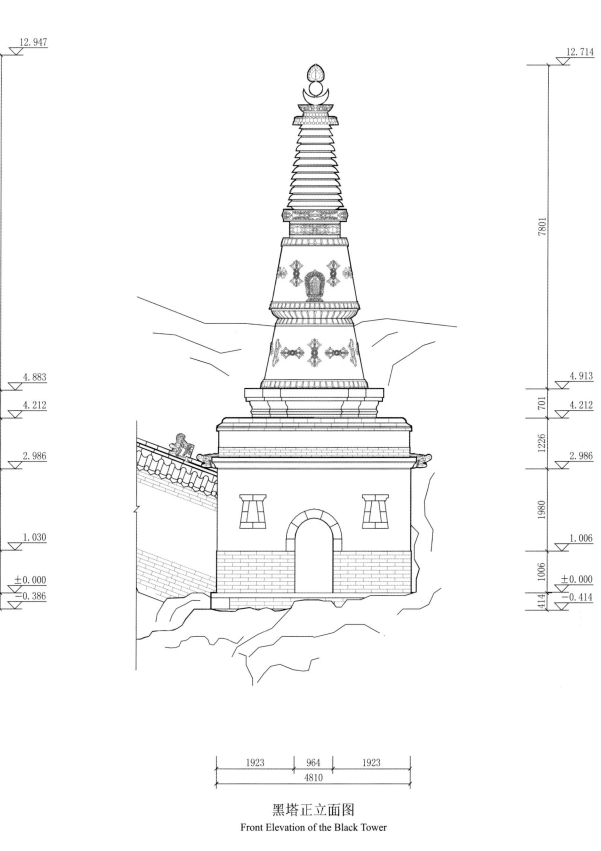

白塔正立面图
Front Elevation of the White Tower

黑塔正立面图
Front Elevation of the Black Tower

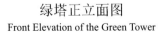

13.883

8623

5.260
720
4.540
1025
3.515

2510

1.005
1005
±0.000
301
−0.301

160 1915 990 1915 160
5140

绿塔正立面图
Front Elevation of the Green Tower

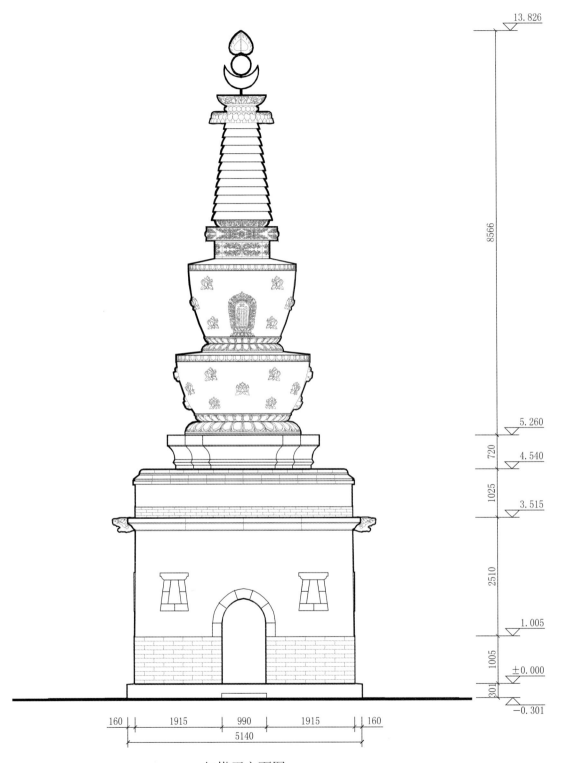

13.826

8566

5.260
720
4.540
1025
3.515

2510

1.005
1005
±0.000
301
−0.301

160 1915 990 1915 160
5140

红塔正立面图
Front Elevation of the Red Tower

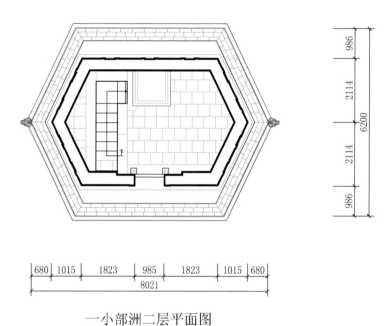

一小部洲二层平面图

Second Floor Plan of the *Yixiaobuzhou*

一小部洲一层平面图

First Floor Plan of the *Yixiaobuzhou*

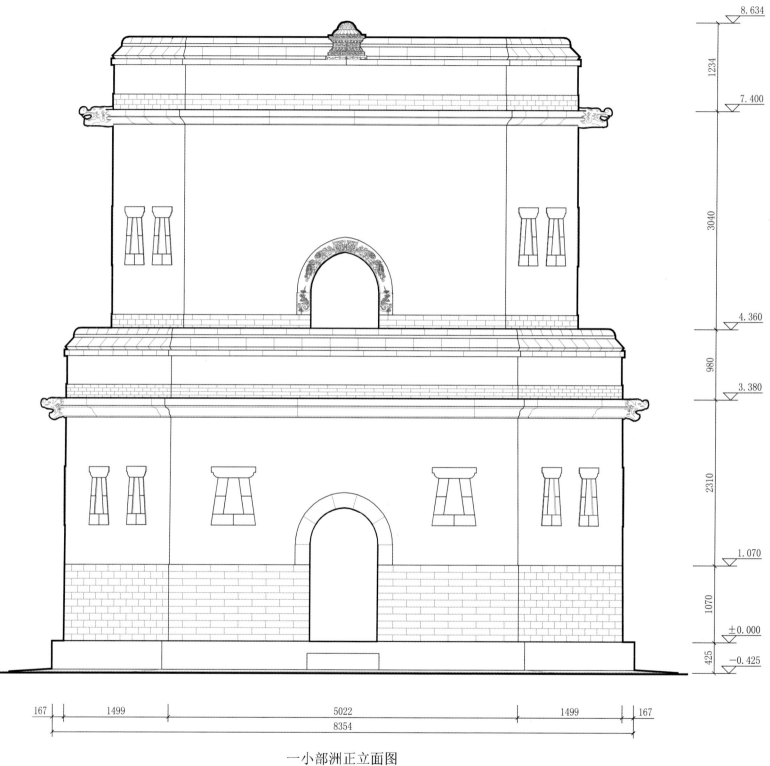

一小部洲正立面图

Front Elevation of the *Yixiaobuzhou*

多宝塔正立面图
Front Elevation of the *Duobao* Tower

598

1706

2902

598

2760 1021 2458 1200 1665

6.344 5.323 2.865 1.665 ±0.000

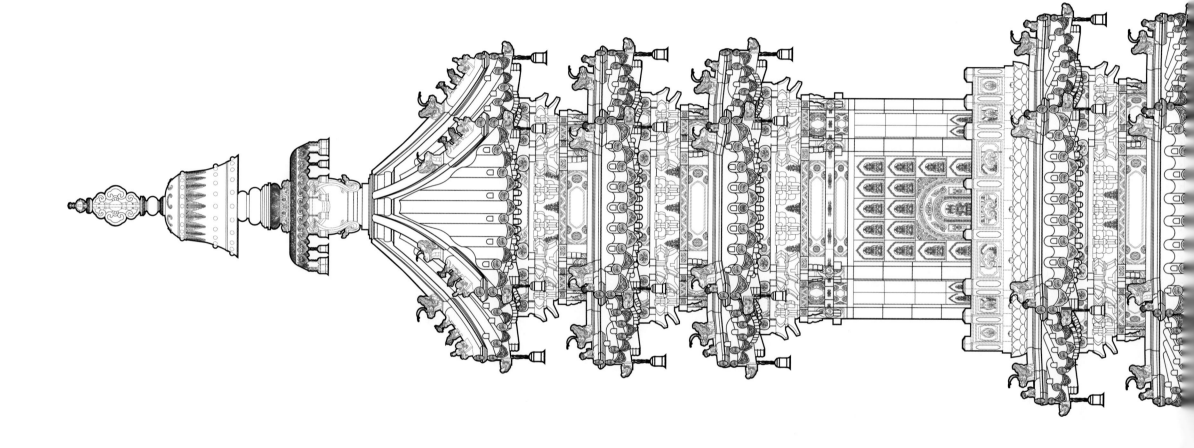

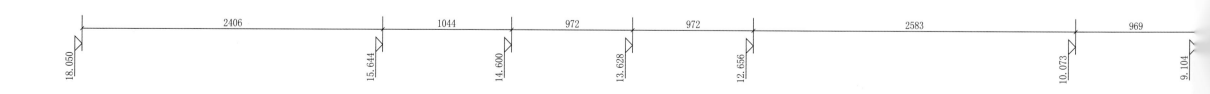

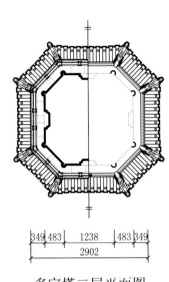

多宝塔二层平面图

Second Floor Plan of the *Duobao* Tower

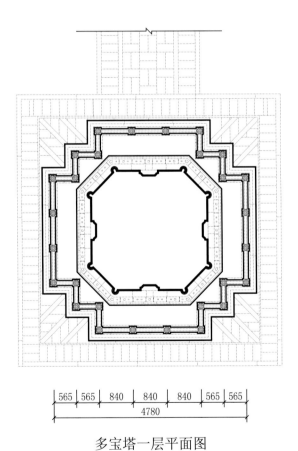

多宝塔一层平面图

First Floor Plan of the *Duobao* Tower

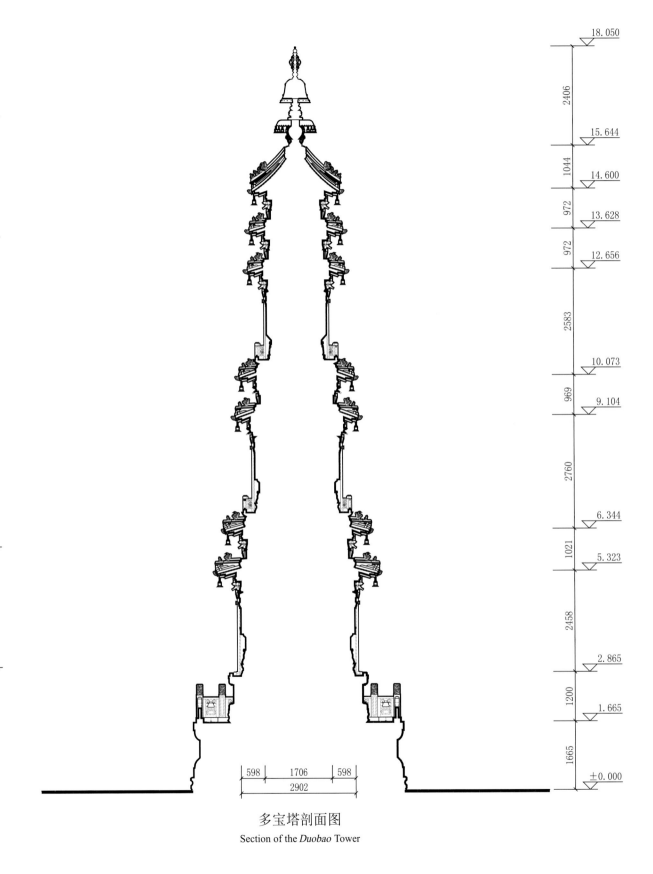

多宝塔剖面图

Section of the *Duobao* Tower

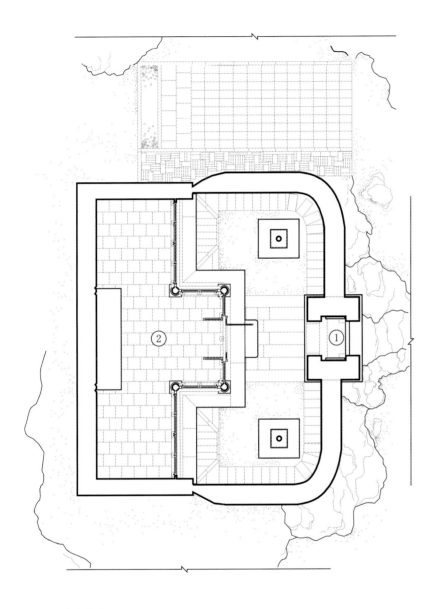

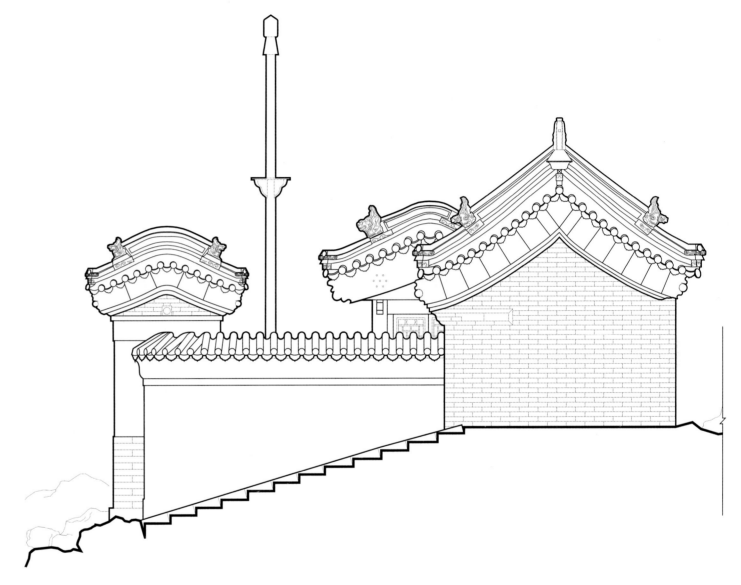

1 山门 Gate
2 正殿 Mail Hall

妙觉寺组群总平面图
Site Plan of the *Miaojue* Temple Complex

0 0.4   2m

妙觉寺组群侧立面图
Side Elevation of the *Miaojue* Temple Complex

0 0.2   1m

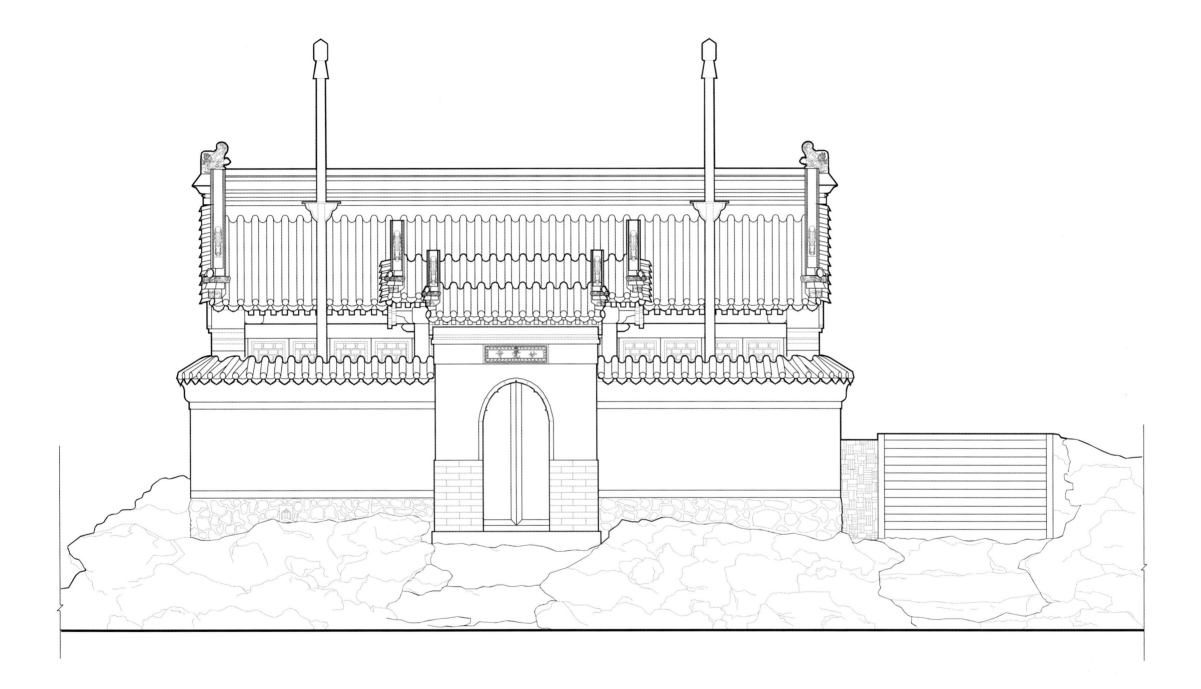

妙觉寺组群正立面图
Front Elevation of the *Miaojue* Temple Complex

0 0.2   1m

中国古建筑测绘大系 · 园林建筑 —— 颐和园（第二版）

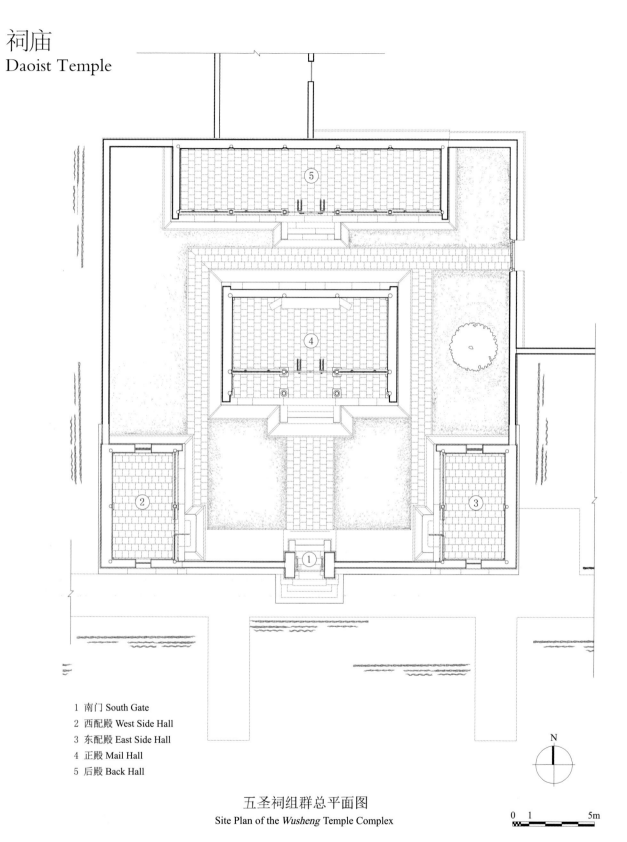

1 南门 South Gate
2 西配殿 West Side Hall
3 东配殿 East Side Hall
4 正殿 Mail Hall
5 后殿 Back Hall

N

五圣祠组群总平面图
Site Plan of the *Wusheng* Temple Complex

0 1   5m

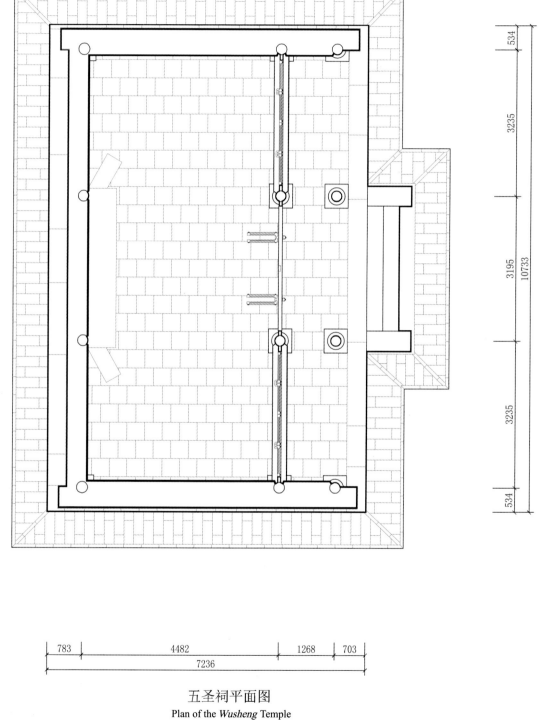

五圣祠平面图
Plan of the *Wusheng* Temple

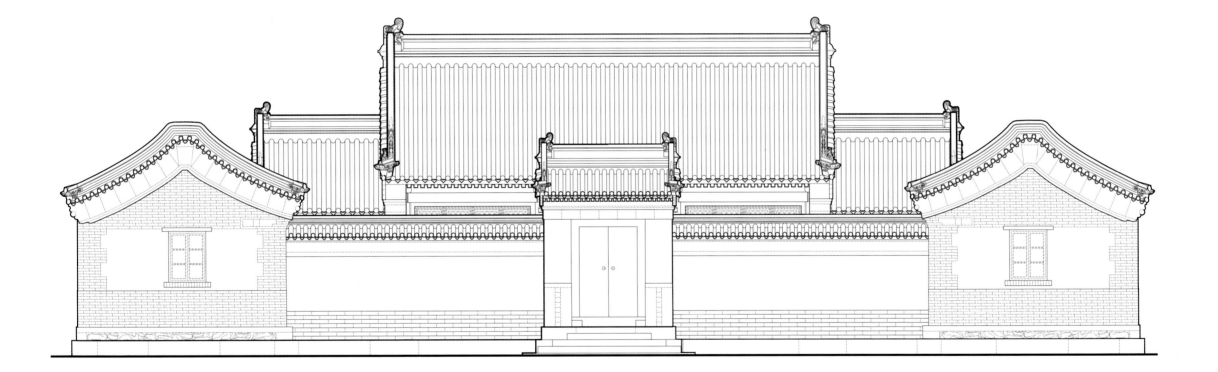

五圣祠组群正立面图

Front Elevation of the *Wusheng* Temple Complex

0 0.4   2m

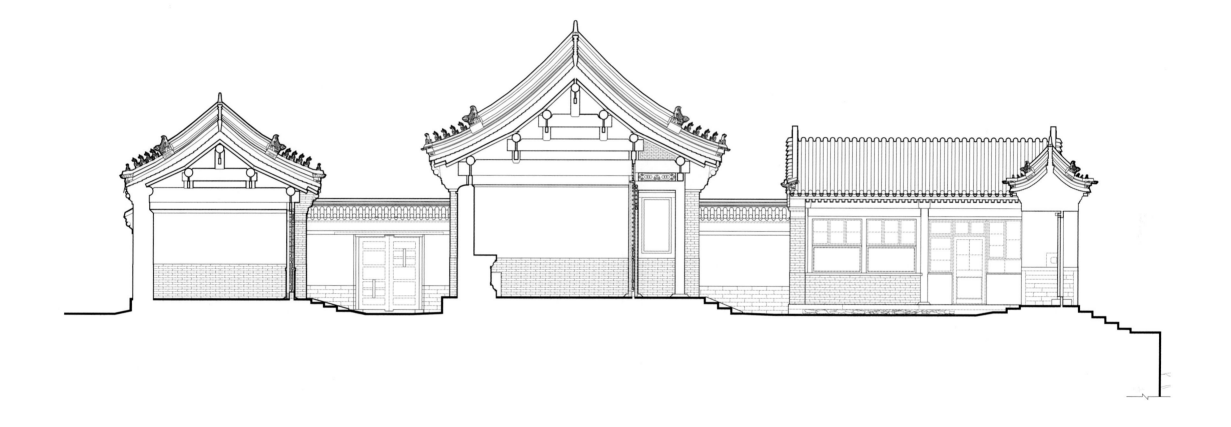

五圣祠组群总剖面图

Section of the *Wusheng* Temple Complex

0 0.4   2m

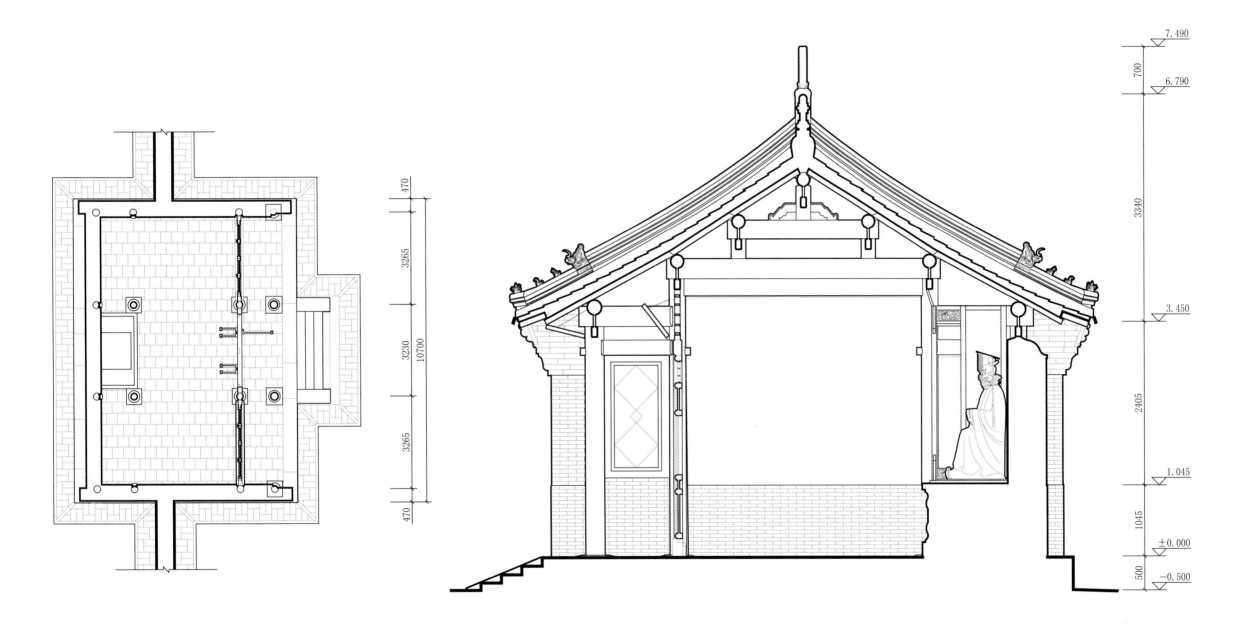

广润灵雨祠正殿平面图

Plan of the Mail Hall of *Guangrunlingyu* Temple

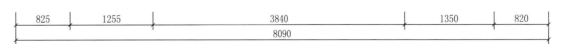

广润灵雨祠正殿明间剖面图

Mingjian Section of the Mail Hall of *Guangrunlingyu* Temple

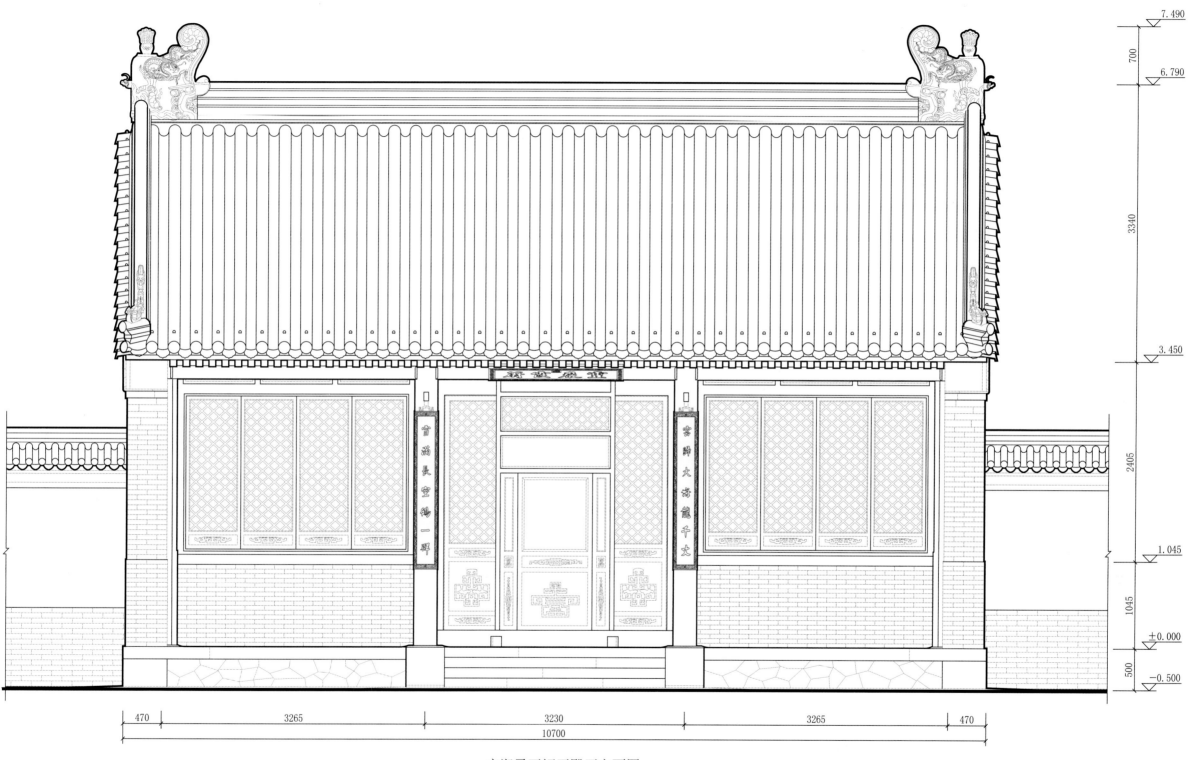

广润灵雨祠正殿正立面图
Front Elevation of the Mail Hall of *Guangrunlingyu* Temple

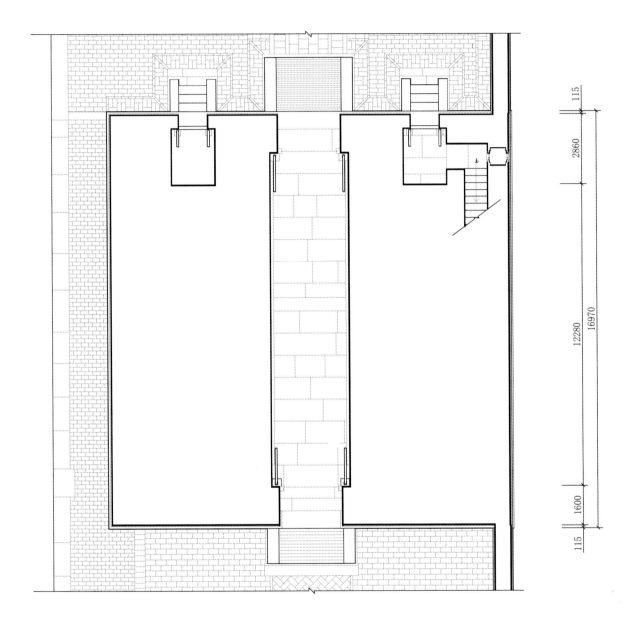

文昌阁一层平面图
First Floor Plan of the *Wenchang* Gate Pavilion

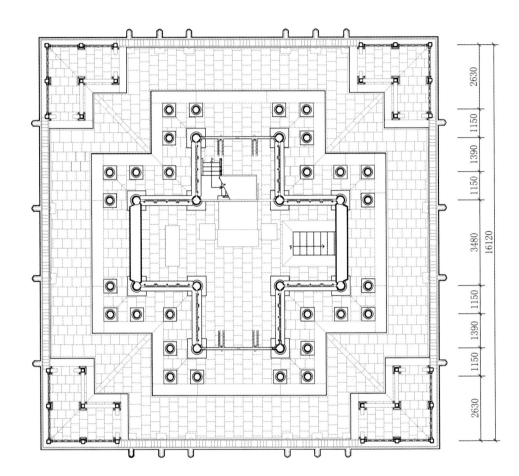

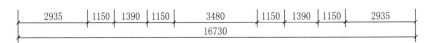

文昌阁二层平面图
Second Floor Plan of the *Wenchang* Gate Pavilion

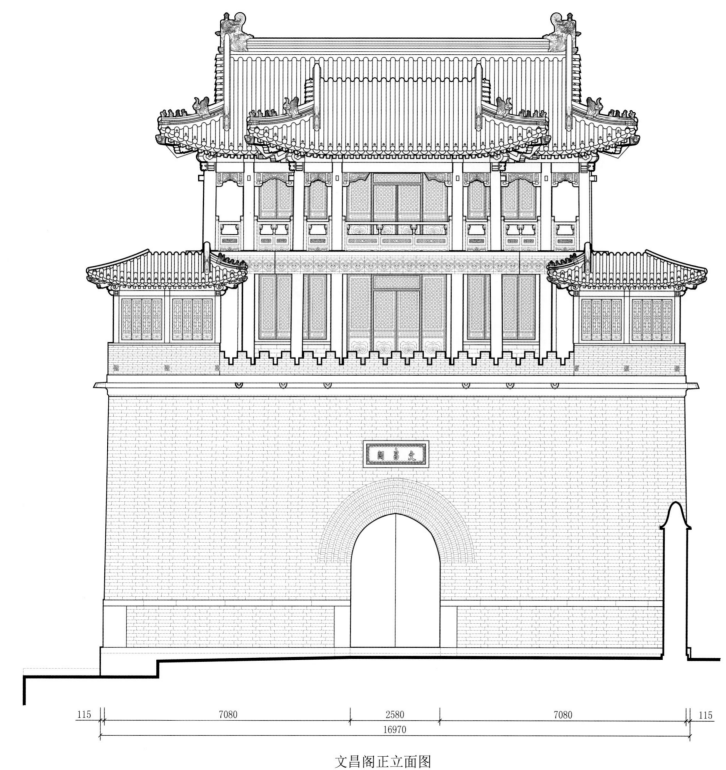

17.659

16.955

704

3330

13.625

2600

11.025

2555

8.470

1110

7.360

7360

±0.000

115　　7080　　2580　　7080　　115

16970

文昌阁正立面图

Front Elevation of the *Wenchang* Gate Pavilion

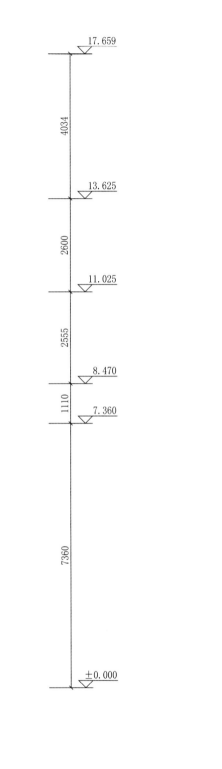

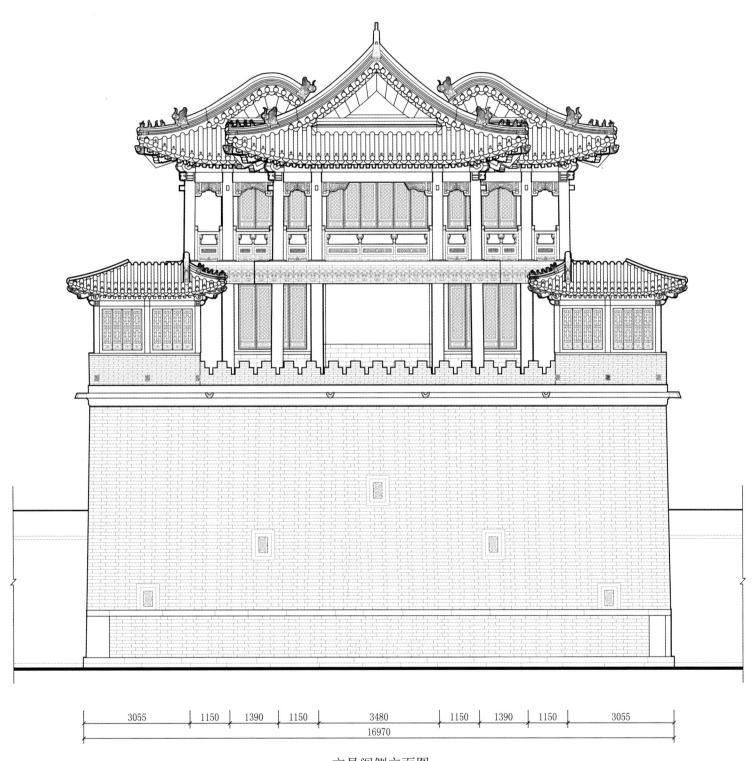

17.659

4034

13.625

2600

11.025

2555

8.470

1110

7.360

7360

±0.000

| 3055 | 1150 | 1390 | 1150 | 3480 | 1150 | 1390 | 1150 | 3055 |

16970

文昌阁侧立面图

Side Elevation of the *Wenchang* Gate Pavilion

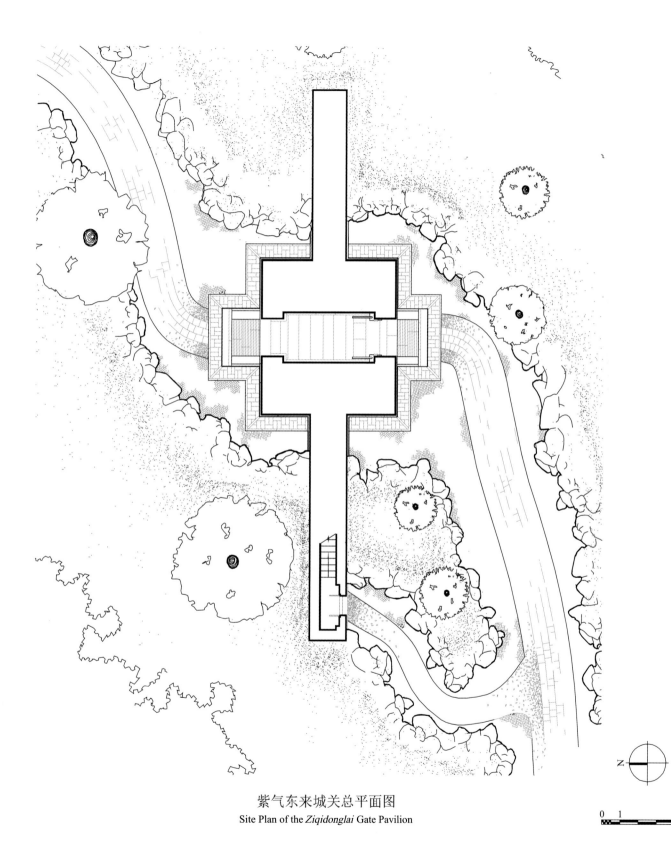

中国古建筑测绘大系·园林建筑——颐和园（第二版）

127

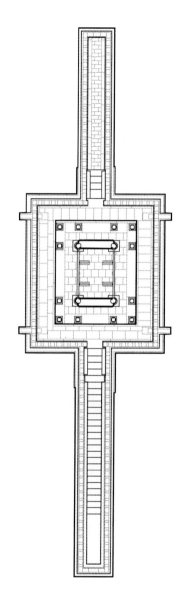

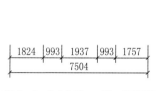

紫气东来城关总平面图
Site Plan of the *Ziqidonglai* Gate Pavilion

0 1 5m

紫气东来城关二层平面图
Second Floor Plan of the *Ziqidonglai* Gate Pavilion

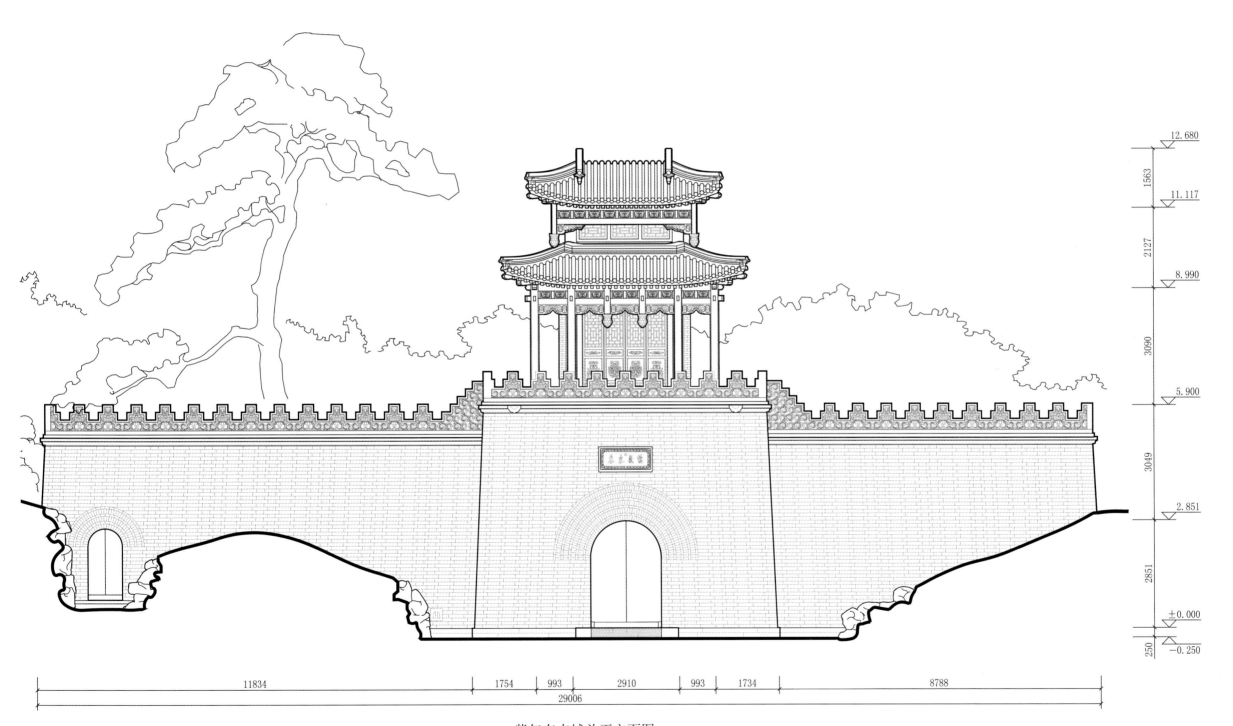

12.680

1563

11.117

2127

8.990

3090

5.900

3049

2.851

2851

±0.000

250 −0.250

11834

1754 993 2910 993 1734

8788

29006

紫气东来城关正立面图
Front Elevation of the *Ziqidonglai* Gate Pavilion

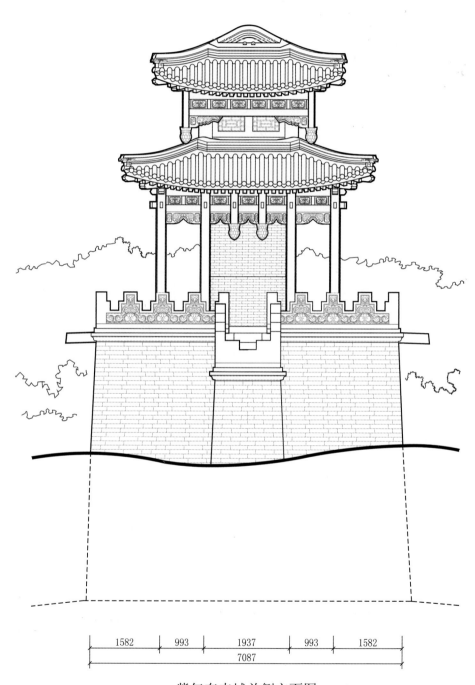

紫气东来城关侧立面图

Side Elevation of the *Ziqidonglai* Gate Pavilion

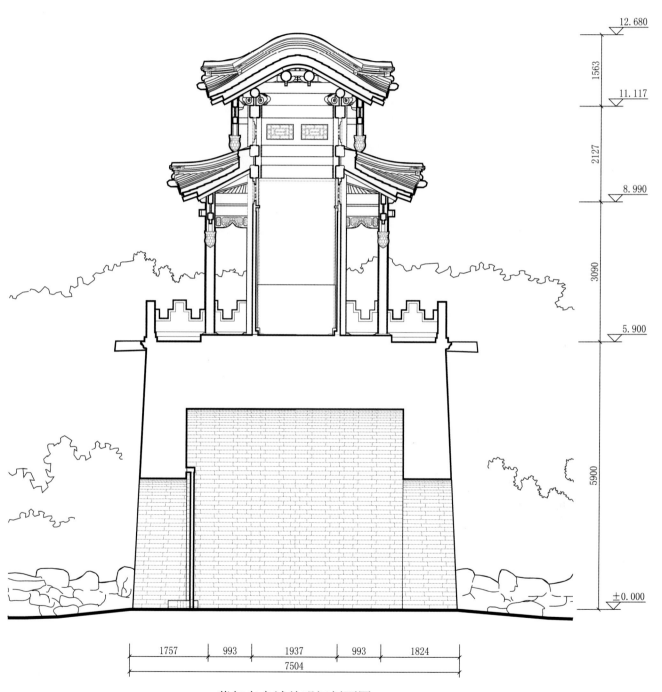

紫气东来城关明间剖面图

Mingjian Section of the *Ziqidonglai* Gate Pavilion

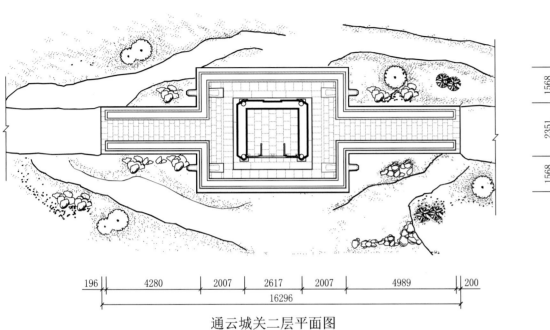

通云城关二层平面图
Second Floor Plan of the *Tongyun* Gate Pavilion

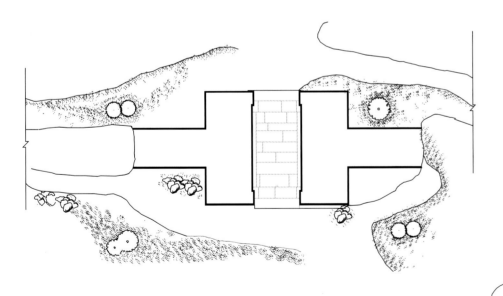

通云城关总平面图
Site Plan of the *Tongyun* Gate Pavilion

0 0.6   3m

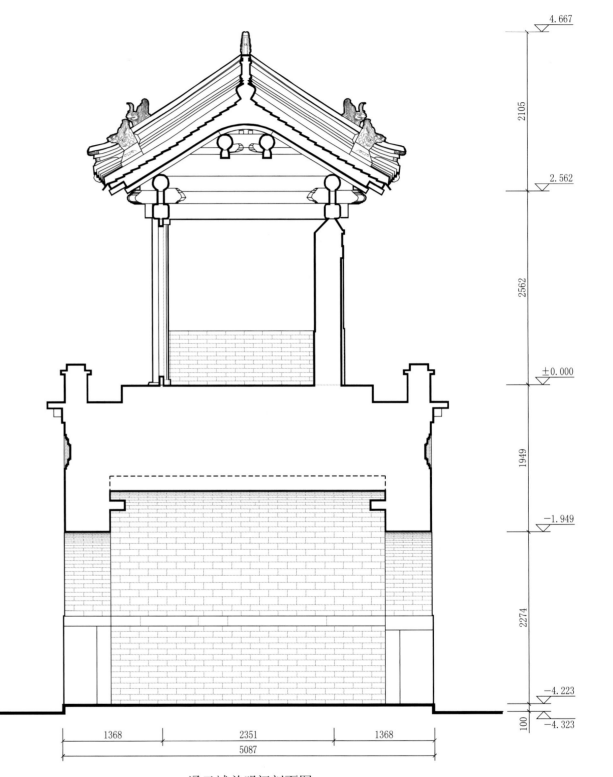

通云城关明间剖面图
Mingjian Section of the *Tongyun* Gate Pavilion

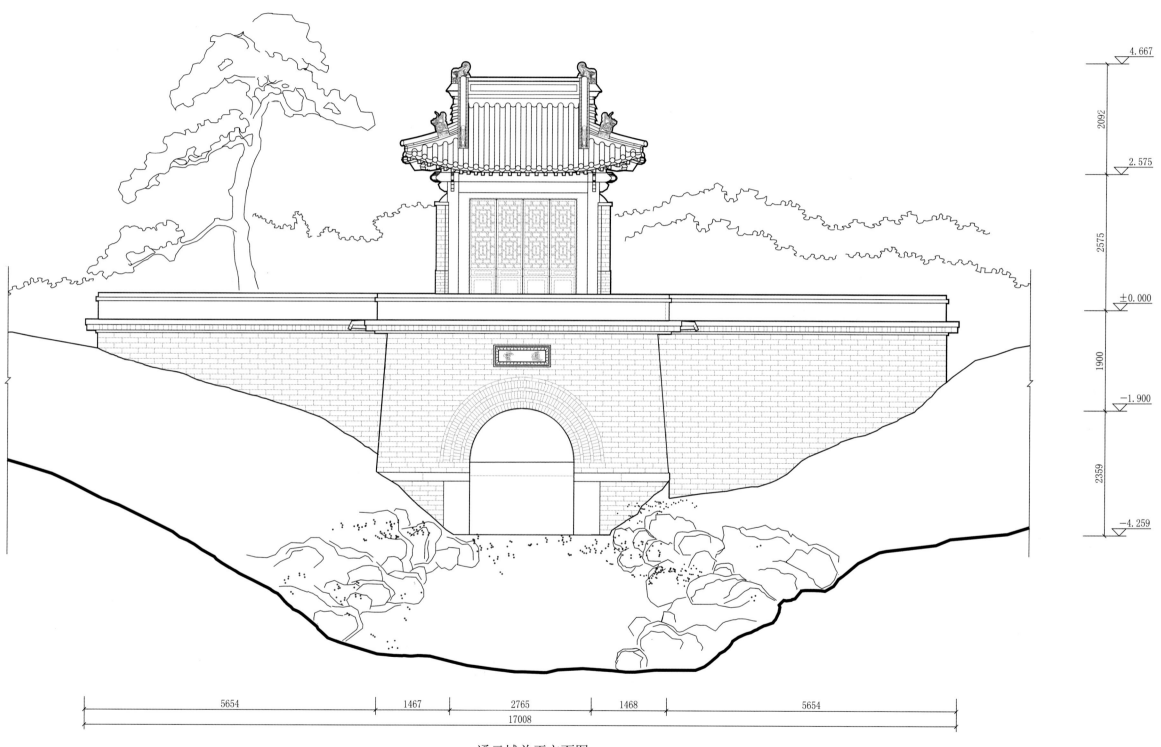

4.667

2092

2.575

2575

±0.000

1900

−1.900

2359

−4.259

匾额

5654　　1467　　2765　　1468　　5654

17008

通云城关正立面图

Front Elevation of the *Tongyun* Gate Pavilion

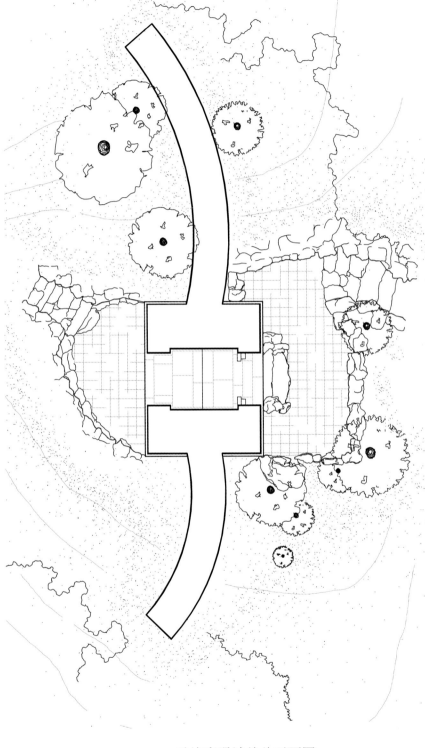

千峰彩翠城关总平面图

Site Plan of the *Qianfengcaicui* Gate Pavilion

0 0.6   3m

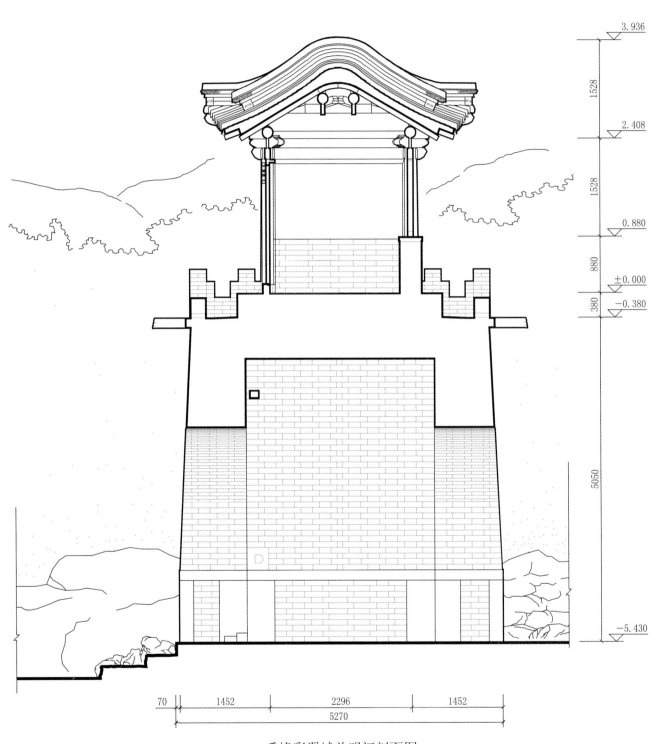

千峰彩翠城关明间剖面图

Mingjian Section of the *Qianfengcaicui* Gate Pavilion

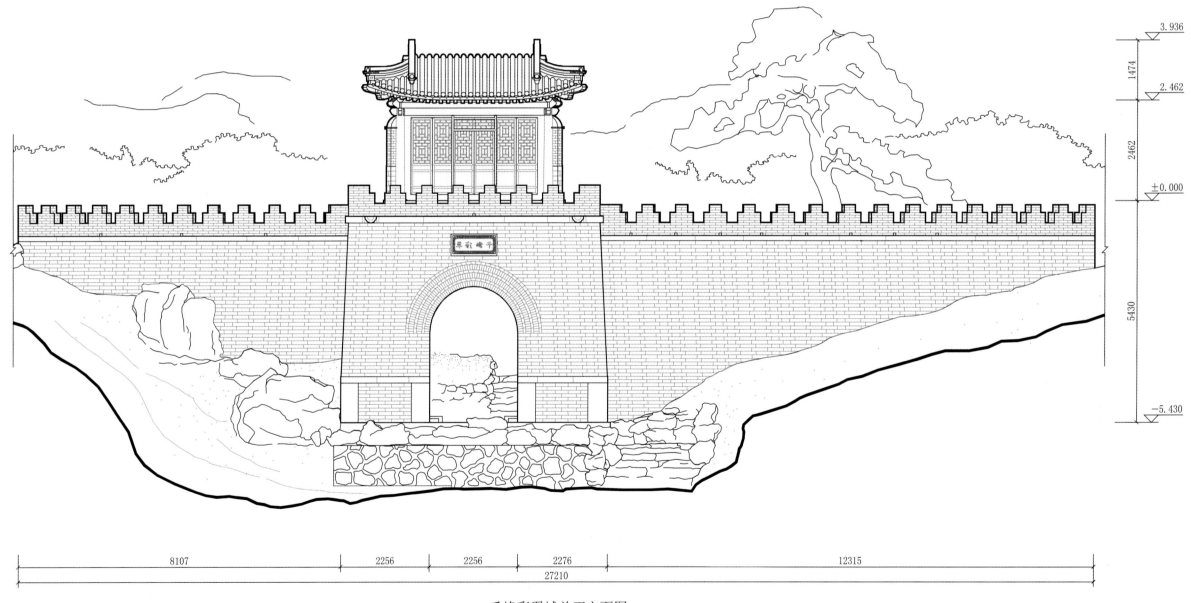

千峰彩翠城关正立面图
Front Elevation of the *Qianfengcaicui* Gate Pavilion

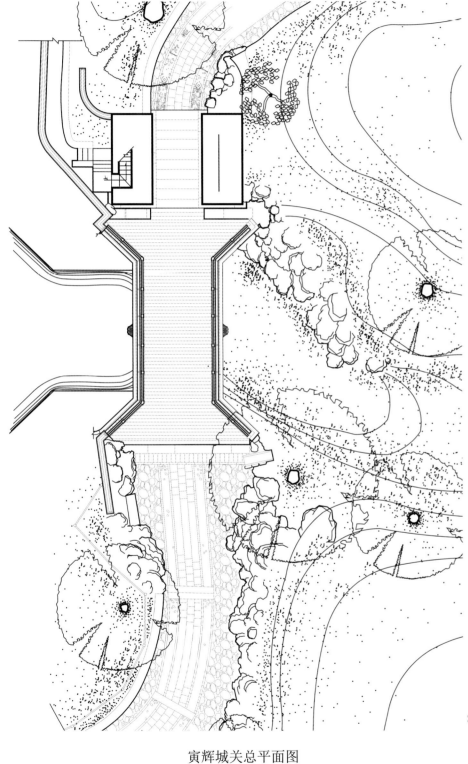

寅辉城关总平面图

Site Plan of the *Yinhui* Gate Pavilion

0 0.6   3m

9.553

2037

7.516

2796

4.720

3423

1.297

1297

±0.000

130

−0.130

165   2287   2624   2287   165

7528

寅辉城关正立面图

Front Elevation of the *Yinhui* Gate Pavilion

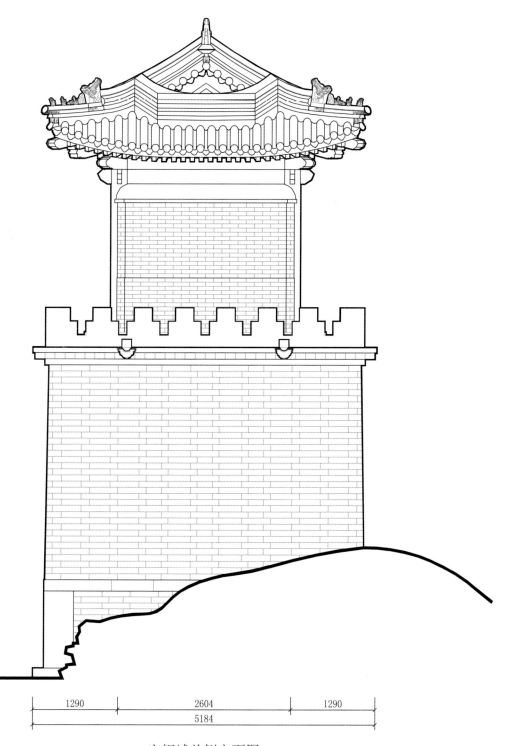

寅辉城关侧立面图

Site Plan of the *Yinhui* Gate Pavilion

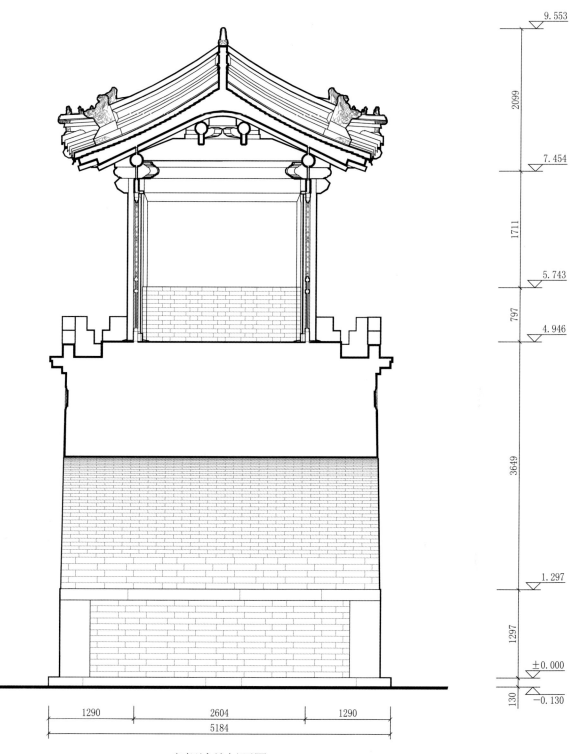

寅辉城关剖面图

Section of the *Yinhui* Gate Pavilion

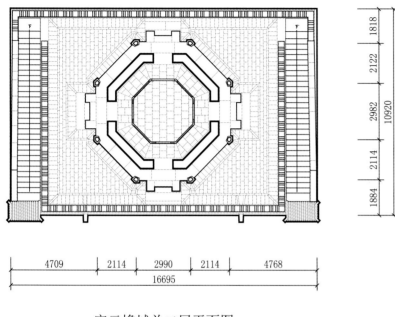

19.841

4122

15.719

3180

12.539

4310

8.229

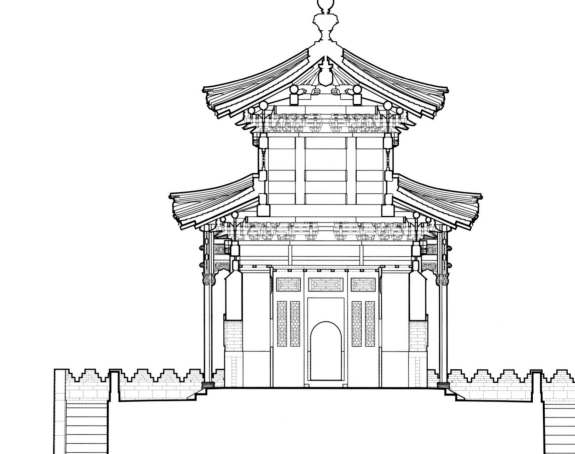

1818
2122
2982
10920
2114
1884

4709 | 2114 | 2990 | 2114 | 4768
16695

宿云檐城关二层平面图
Second Floor Plan of the *Suyunyan* Gate Pavilion

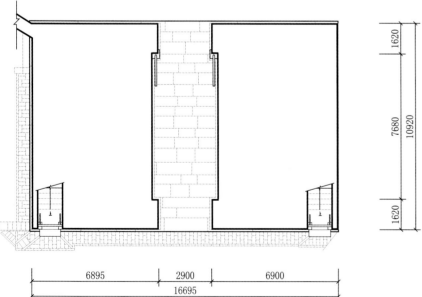

1620

7680
10920

1620

6895 | 2900 | 6900
16695

宿云檐城关一层平面图
First Floor Plan of the *Suyunyan* Gate Pavilion

8229

±0.000

4736 | 7218 | 4741
16695

宿云檐城关剖面图
Section of the *Suyunyan* Gate Pavilion

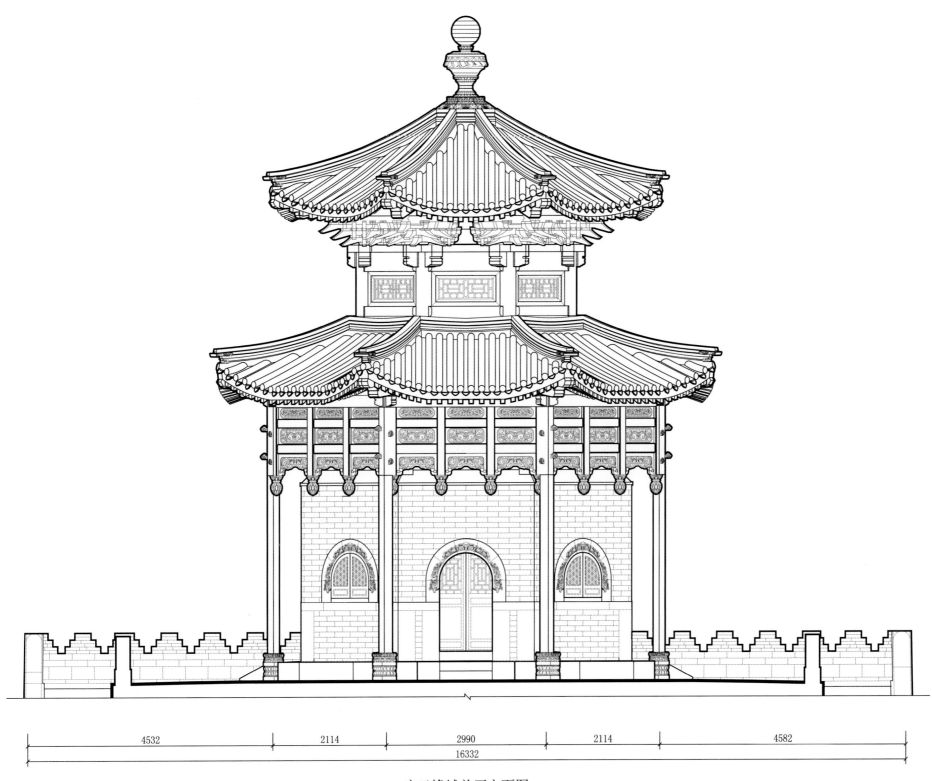

宿云檐城关正立面图
Front Elevation of the *Suyunyan* Gate Pavilion

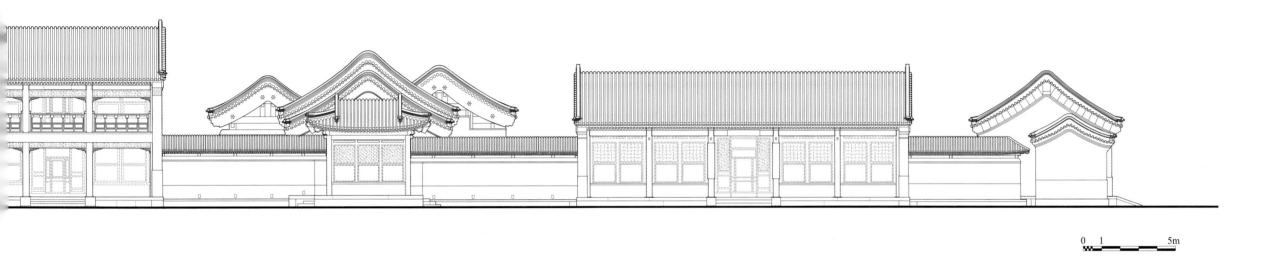

0 1 5m

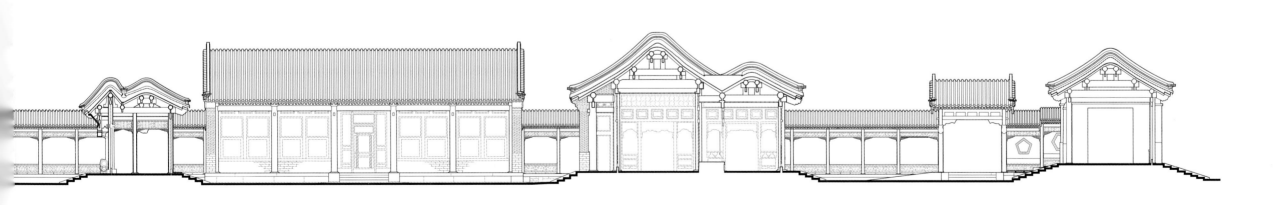

0 1 5m

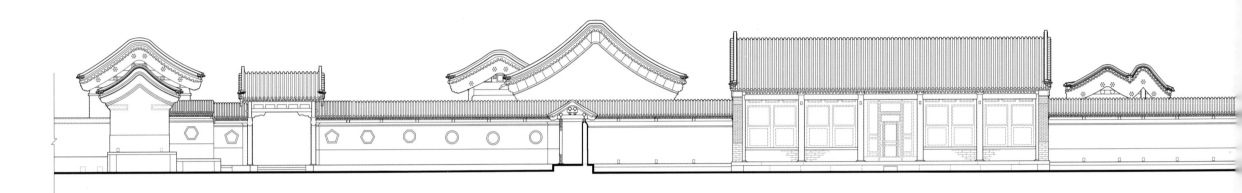

玉澜堂宜芸馆组群总立面图

Elevation of the *Yulan* Hall and *Yiyun* Hall Complex

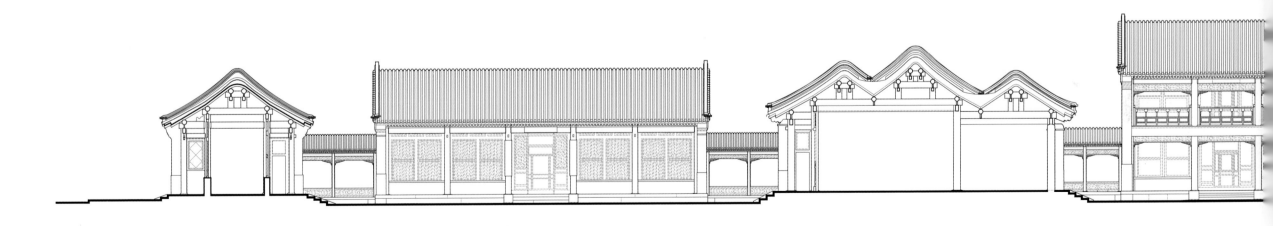

玉澜堂宜芸馆组群总剖面图

Section of the *Yulan* Hall and *Yiyun* Hall Complex

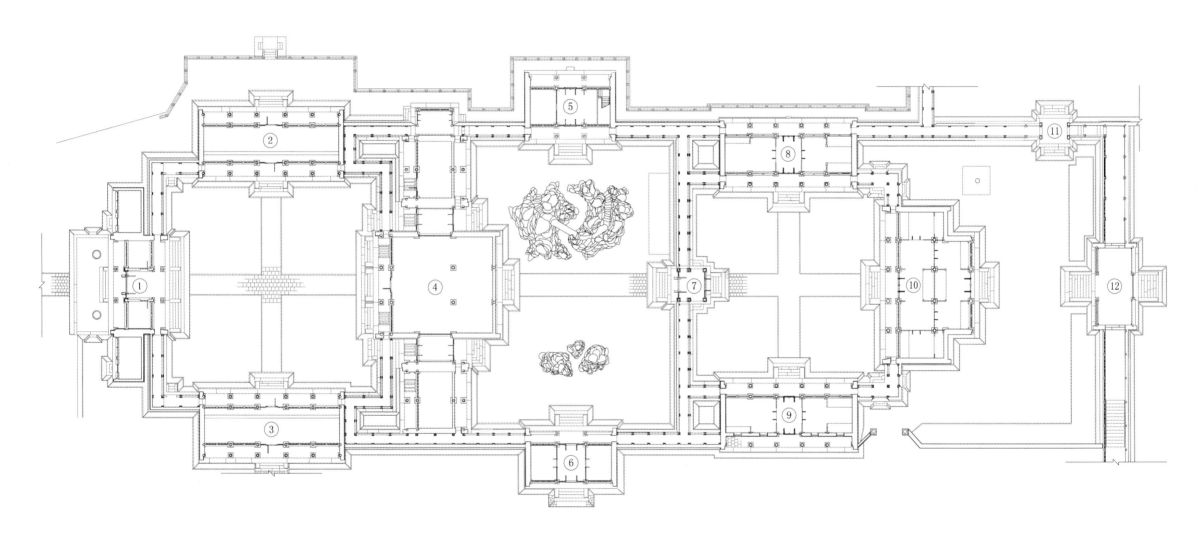

1　玉澜门 *Yulan* Gate　　2　藕香榭 *Ouxiang* Hall　　3　霞芬室 *Xiafen* Hall　　4　玉澜堂 *Yulan* Hall　　5　夕佳楼 *Xijia* Pavilion　　6　穿堂门 Passage Door

7　宜芸门 *Yiyun* Gate　　8　近西轩 *Jinxi* Pavilion　　9　道存斋 *Daocun* Studio　　10　宜芸馆 *Yiyun* Hall　　11　宜芸馆垂花门 Festooned Door of *Yiyun* Hall　　12　宜芸馆后宫门 Back Gate of *Yiyun* Hall

玉澜堂宜芸馆组群总平面图

Site Plan of the *Yulan* Hall and *Yiyun* Hall Complex

0　1　　5　　10m

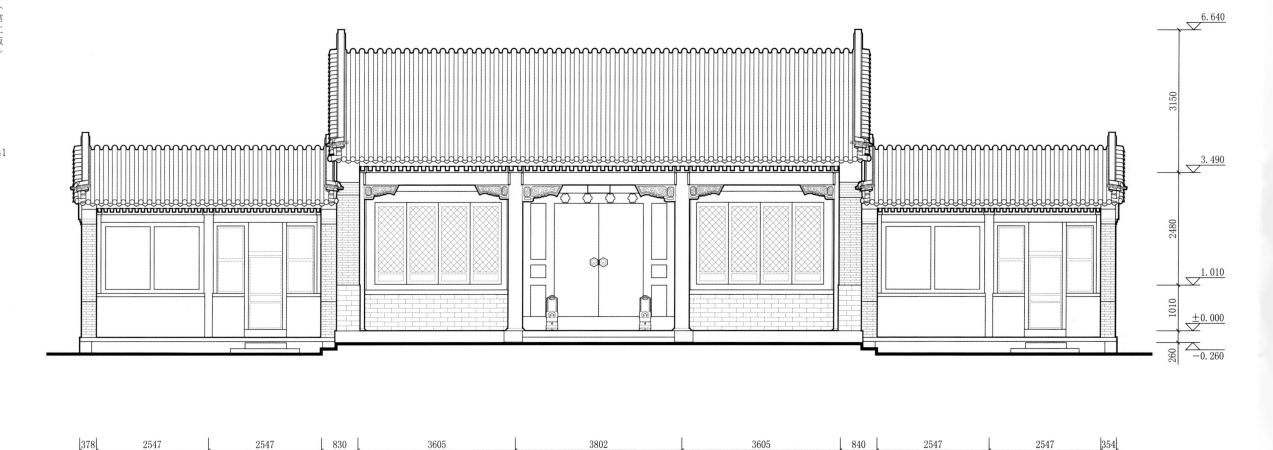

玉澜门正立面图

Front Elevation of the *Yulan* Gate

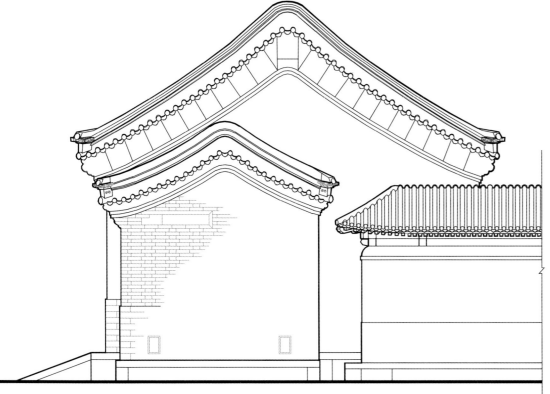

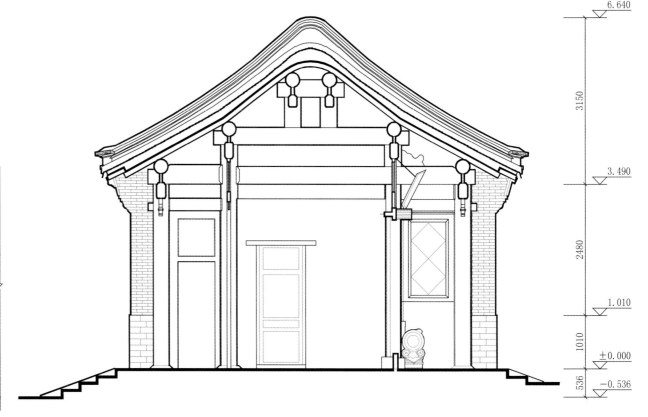

玉澜门侧立面图
Side Elevation of the *Yulan* Gate

玉澜门明间剖面图
Mingjian Section of the *Yulan* Gate

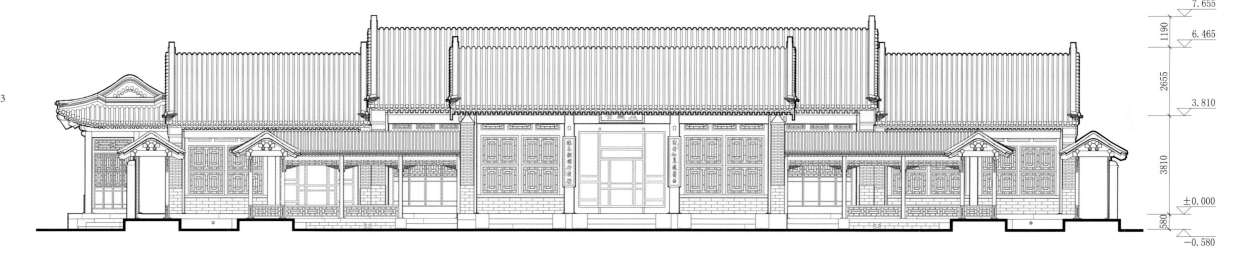

玉澜堂正立面图

Front Elevation of the *Yulan* Hall

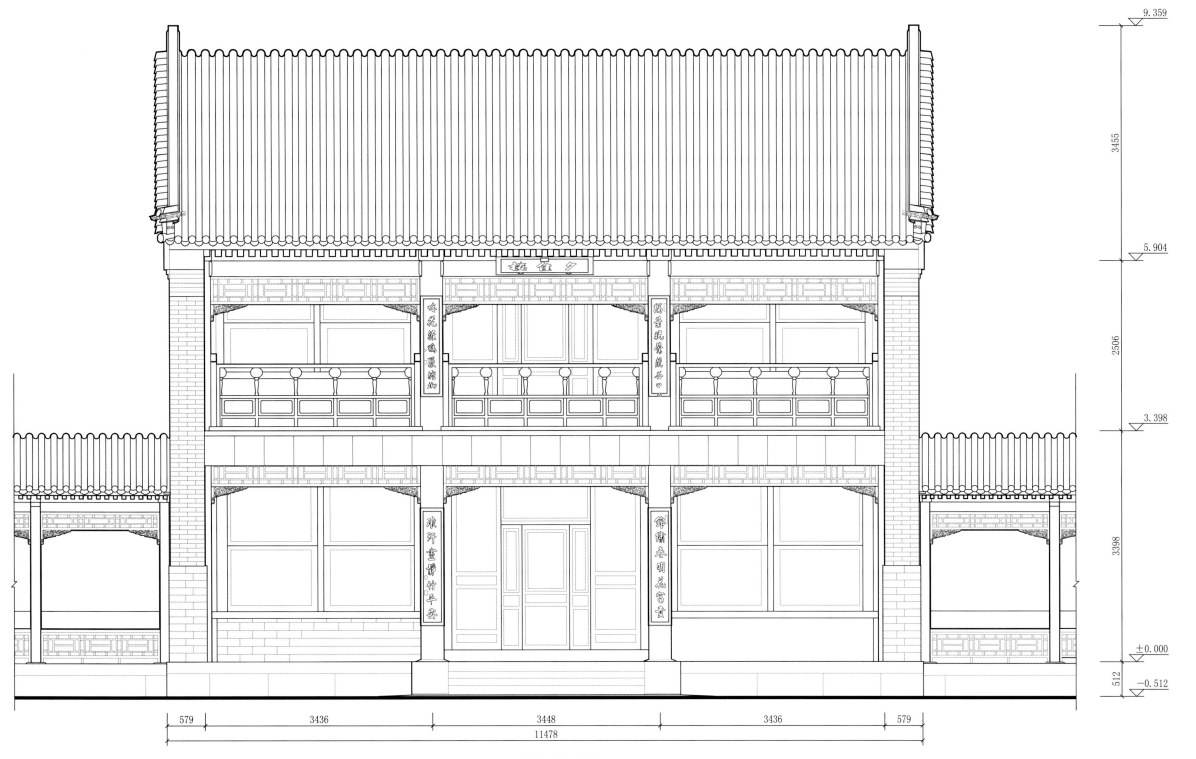

9.359

3455

5.904

2506

144

3.398

3398

±0.000

512 −0.512

579 3436 3448 3436 579

11478

夕佳楼正立面图

Front Elevation of the *Xijia* Pavilion

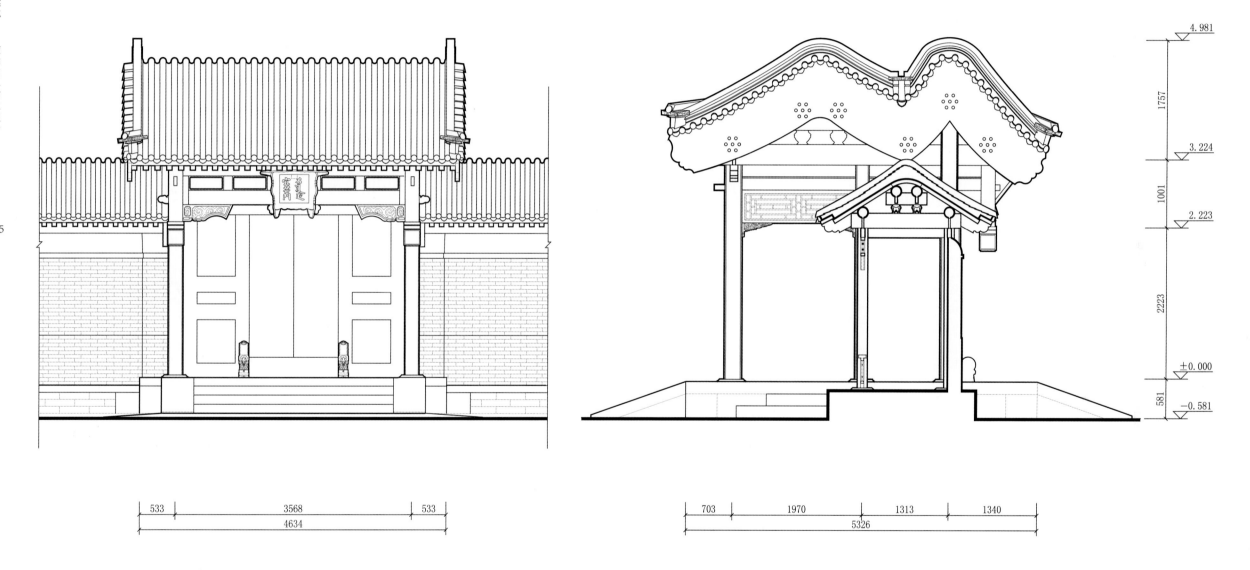

533　　　3568　　　533

4634

703　　1970　　1313　　1340

5326

4.981

1757

3.224

1001

2.223

2223

±0.000

581　　−0.581

宜芸门正立面图

Front Elevation of the *Yiyun* Gate

宜芸门侧立面图

Side Elevation of the *Yiyun* Gate

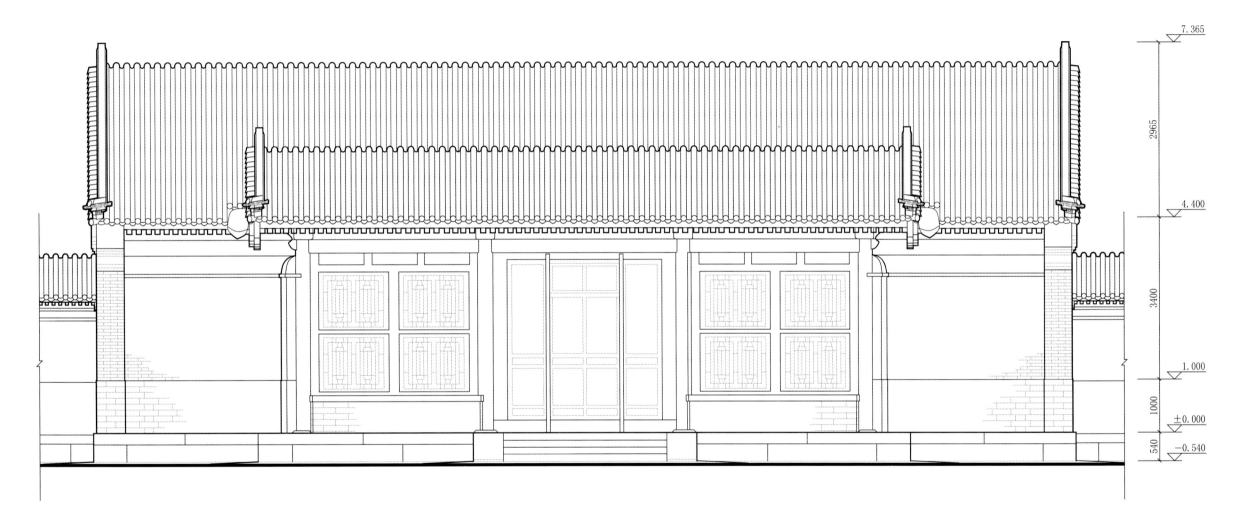

7.365

2965

4.400

3400

1.000

1000

±0.000

540

−0.540

523 | 3570 | 3560 | 3865 | 3560 | 3570 | 523

19171

宜芸馆背立面图

Back Elevation of the *Yiyun* Hall

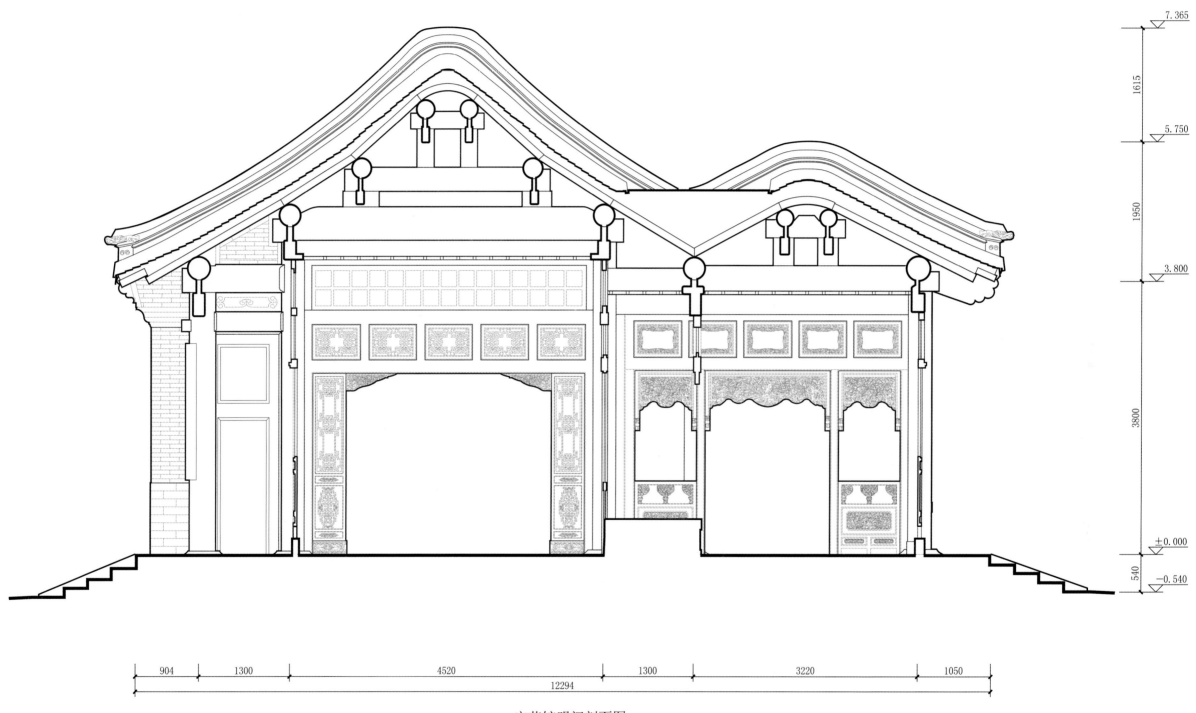

7.365

1615

5.750

1950

3.800

3800

±0.000

540   −0.540

904   1300   4520   1300   3220   1050
12294

宜芸馆明间剖面图

Mingjian Section of the *Yiyun* Hall

1 西一所怀仁憬集门 *Huairenjingji* Gate of *Xiyisuo*
2 西一所东耳房 East Wing Room of *Xiyisuo*
3 西一所照壁 Screen Wall of *Xiyisuo*
4 西一所西配房 West Side Hall of *Xiyisuo*
5 西一所东配房 East Side Hall of *Xiyisuo*
6 西一所正房 Main Hall of *Xiyisuo*
7 西三所 *Xisansuo*　8 西二所 *Xi'ersuo*　9 西四所 *Xisisuo*
10 西五所西厢房 West Side Hall of *Xiwusuo*
11 西五所东厢房 East Side Hall of *Xiwusuo*
12 西五所正房 Main Hall of *Xiwusuo*

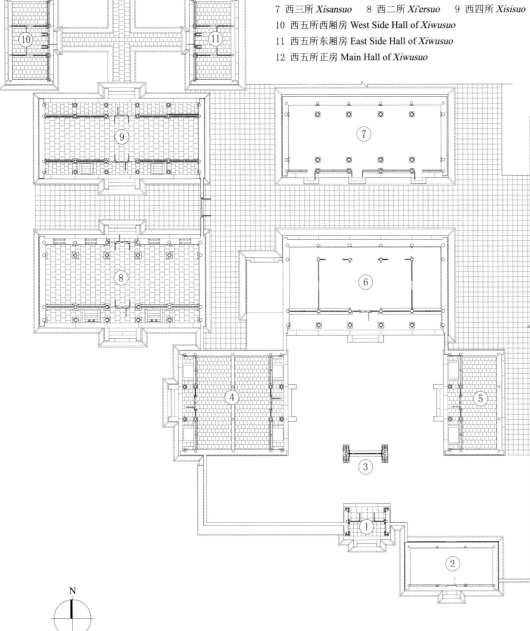

0　1　5　10m

西四所组群总平面图
Site Plan of the *Xisisuo* Complex

7.220

3484

3.736

2701

1.035

1035　±0.000

228　−0.228

965　1380　5070　1380　967
9762

西四所明间剖面图
Mingjian Section of the *Xisisuo*

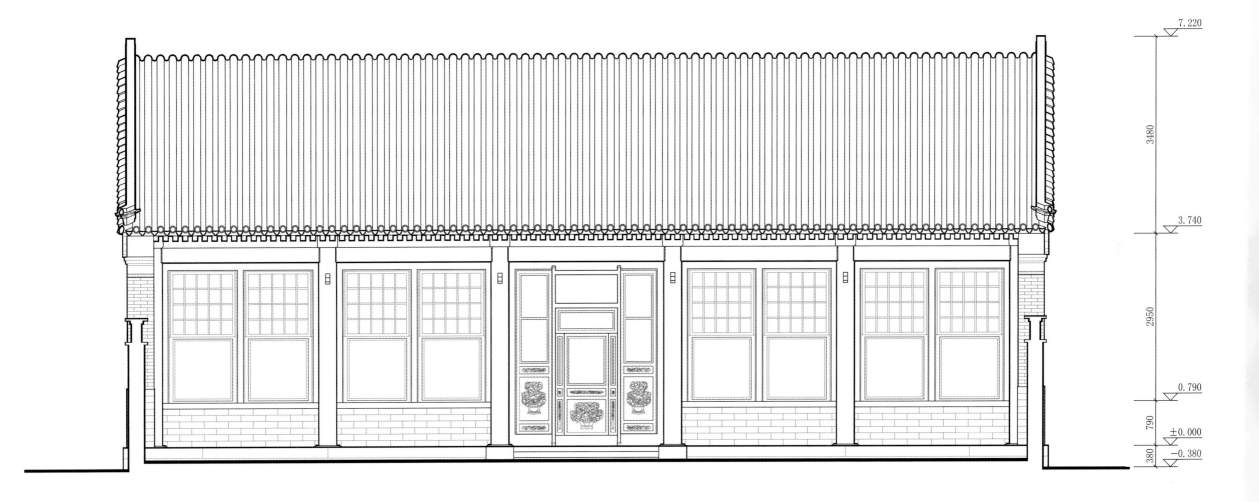

7.220

3480

3.740

2950

0.790

790

±0.000

380

−0.380

539   3150   3150   3150   3150   3150   539

16828

西四所正立面图

Front Elevation of the *Xisisuo*

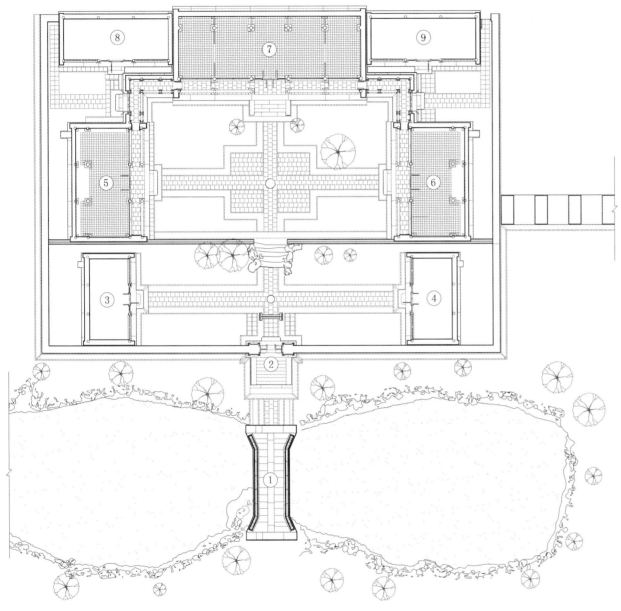

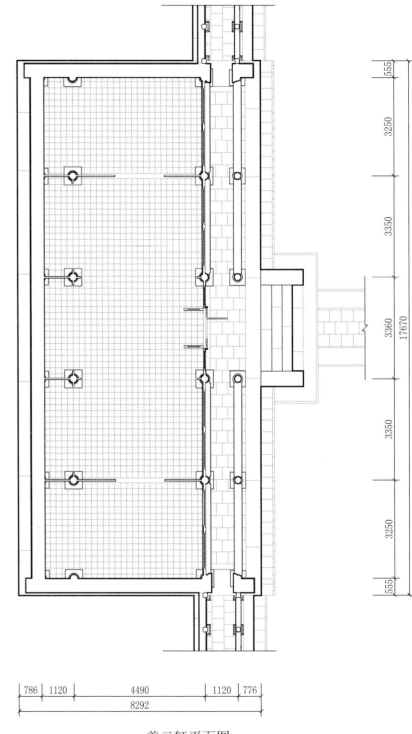

150

1 养云轩桥 Bridge of *Yangyun* Pavilion　2 西洋门 Western Gate　3 西值房 West Duty Room
4 东值房 East Duty Room　5 西配殿 West Side Hall　6 东配殿 East Side Hall
7 养云轩 *Yangyun* Pavilion　8 西耳房 West Wing Room　9 东耳房 East Wing Room

N

养云轩组群总平面图
Site Plan of the *Yangyun* Pavilion Complex

0 1　　5m

养云轩平面图
Plan of the *Yangyun* Pavilion

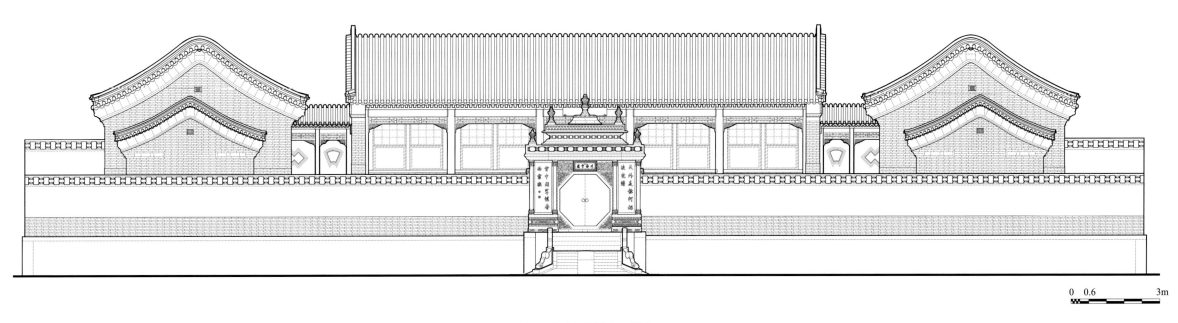

养云轩组群总立面图

Elevation of the *Yangyun* Pavilion Complex

0   0.6   3m

养云轩组群总剖面图

Section of the *Yangyun* Pavilion Complex

0   0.6   3m

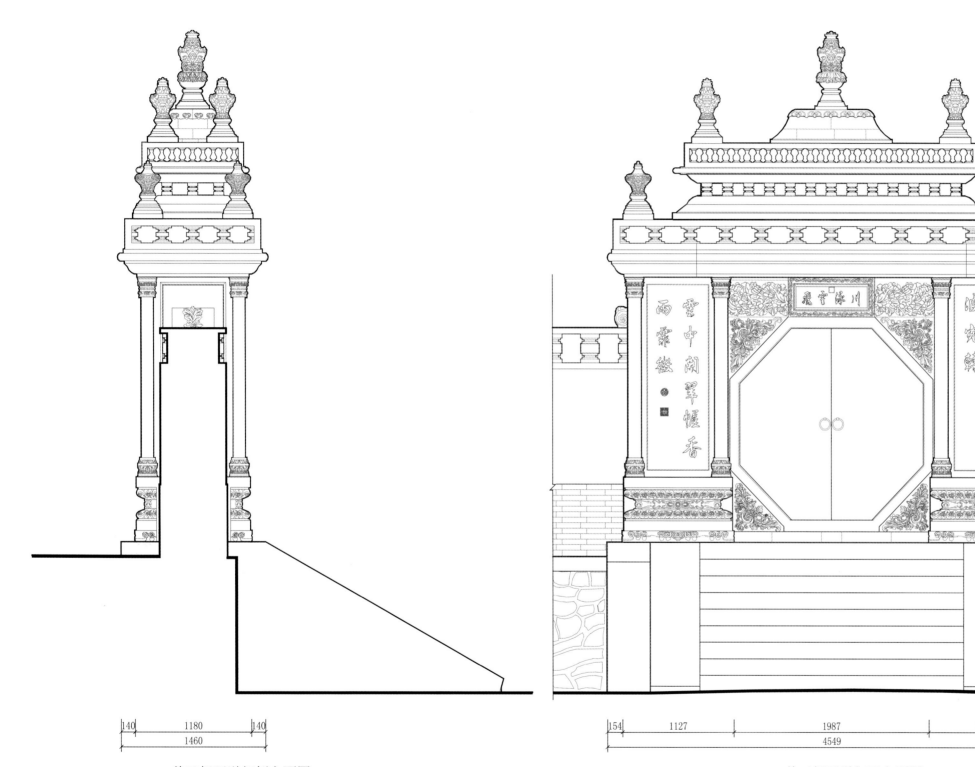

养云轩西洋门侧立面图

Side Elevation of the Western Gate of *Yangyun* Pavilion

养云轩西洋门正立面图

Front Elevation of the Western Gate of *Yangyun* Pavilion

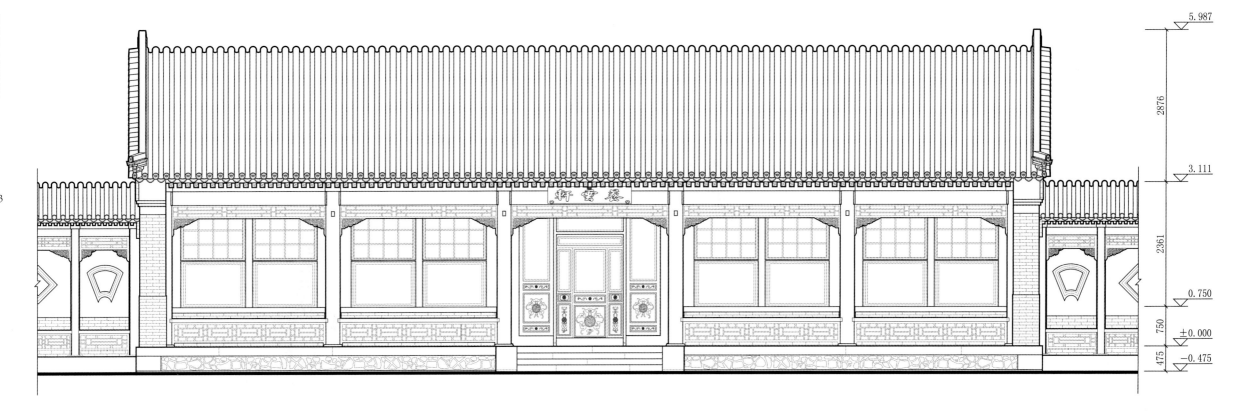

养云轩正立面图

Front Elevation of the *Yangyun* Pavilion

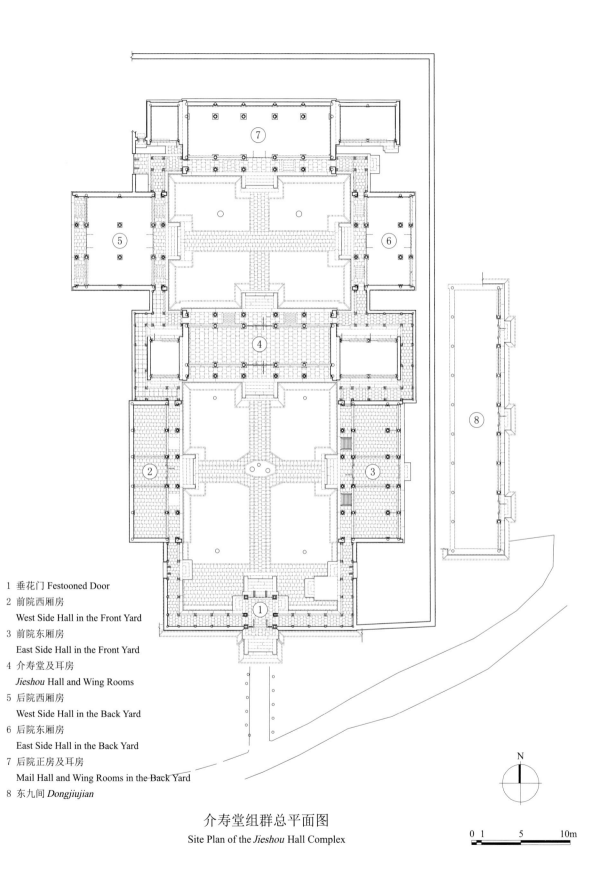

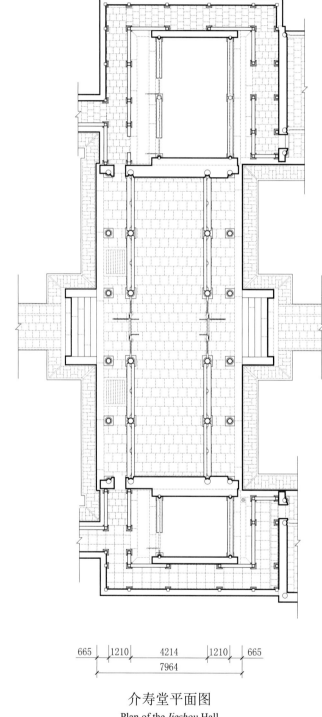

1 垂花门 Festooned Door
2 前院西厢房
  West Side Hall in the Front Yard
3 前院东厢房
  East Side Hall in the Front Yard
4 介寿堂及耳房
  Jieshou Hall and Wing Rooms
5 后院西厢房
  West Side Hall in the Back Yard
6 后院东厢房
  East Side Hall in the Back Yard
7 后院正房及耳房
  Mail Hall and Wing Rooms in the Back Yard
8 东九间 *Dongjiujian*

介寿堂组群总平面图
Site Plan of the *Jieshou* Hall Complex

0 1   5   10m

N

介寿堂平面图
Plan of the *Jieshou* Hall

665 | 1210 | 4214 | 1210 | 665
7964

3190
3174
3586
3174
3190
16314

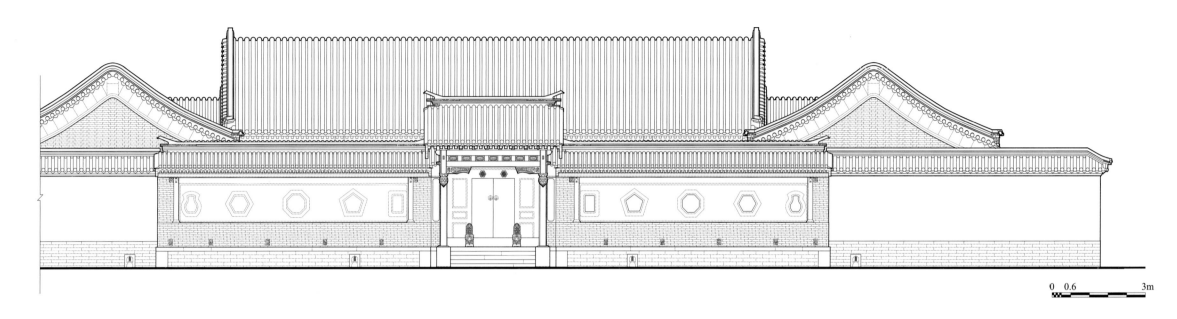

介寿堂组群总立面图

Elevation of the *Jieshou* Hall Complex

介寿堂组群总剖面图

Section of the *Jieshou* Hall Complex

介寿堂正立面图

Front Elevation of the *Jieshou* Hall

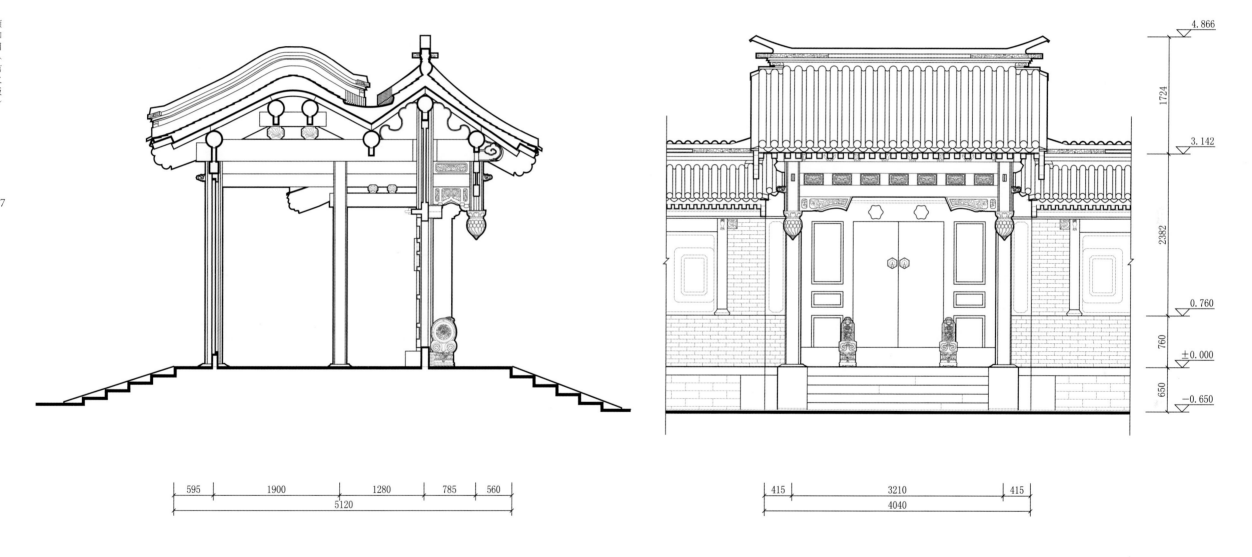

介寿堂垂花门剖面图
Section of the Festooned Door of *Jieshou* Hall

介寿堂垂花门正立面图
Front Elevation of the Festooned Door of *Jieshou* Hall

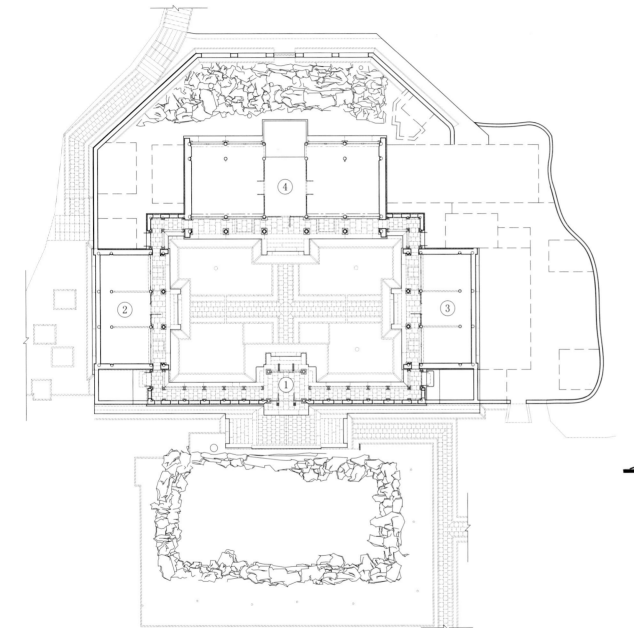

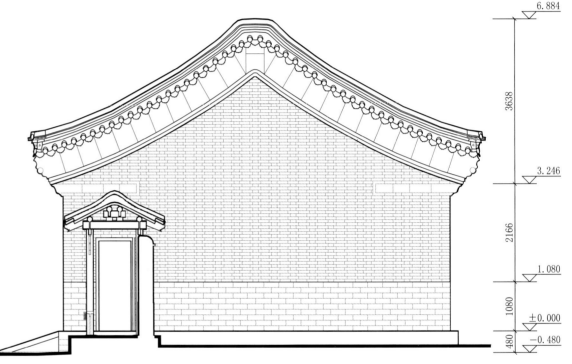

158

1　垂花门 Festooned Door　2　西厢房 West Side Hall
3　东厢房 East Side Hall　4　无尽意轩 *Wujinyi* Pavilion

无尽意轩组群总平面图
Site Plan of the *Wujinyi* Pavilion Complex

0　1　　　5m

无尽意轩侧立面图
Side Elevation of the *Wujinyi* Pavilion

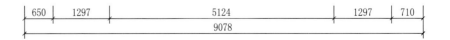

无尽意轩正立面图

Front Elevation of the *Wujinyi* Pavilion

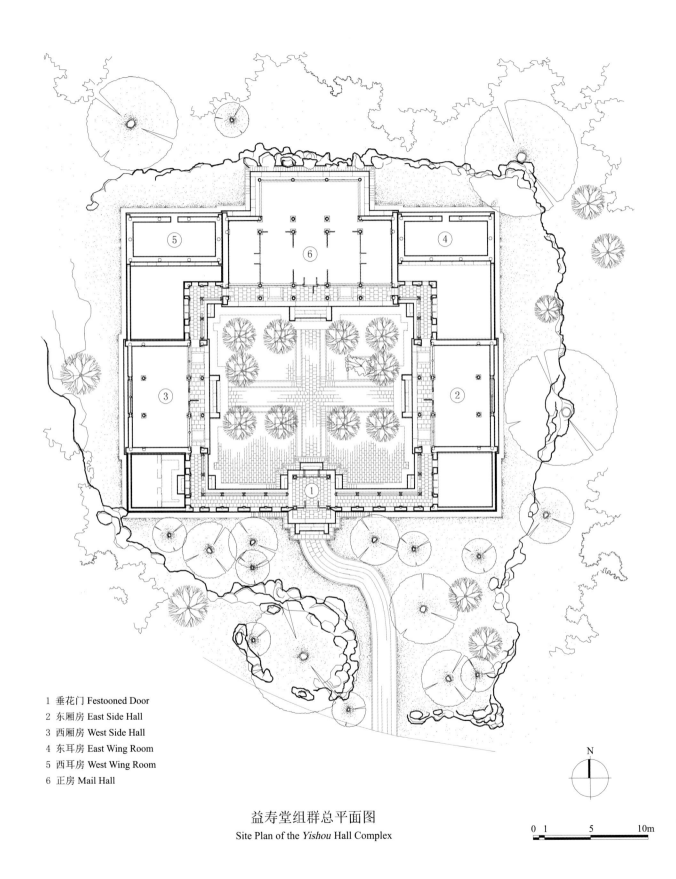

1 垂花门 Festooned Door
2 东厢房 East Side Hall
3 西厢房 West Side Hall
4 东耳房 East Wing Room
5 西耳房 West Wing Room
6 正房 Mail Hall

益寿堂组群总平面图
Site Plan of the *Yishou* Hall Complex

0 1 5 10m

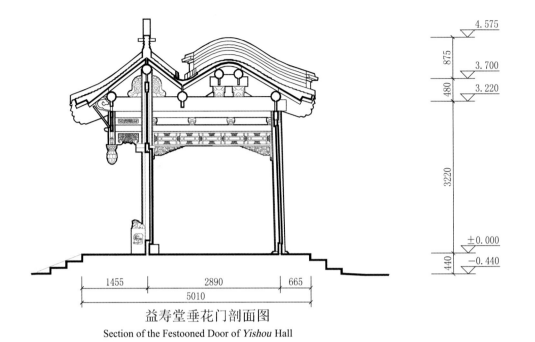

益寿堂垂花门剖面图
Section of the Festooned Door of *Yishou* Hall

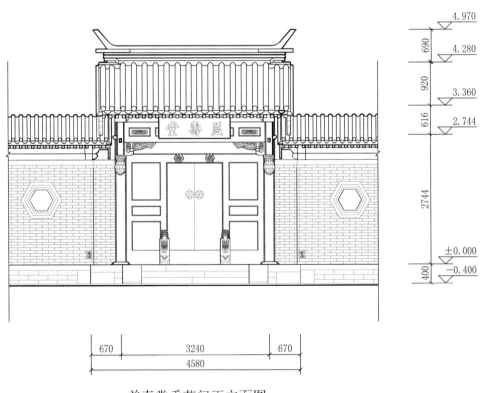

益寿堂垂花门正立面图
Front Elevation of the Festooned Door of *Yishou* Hall

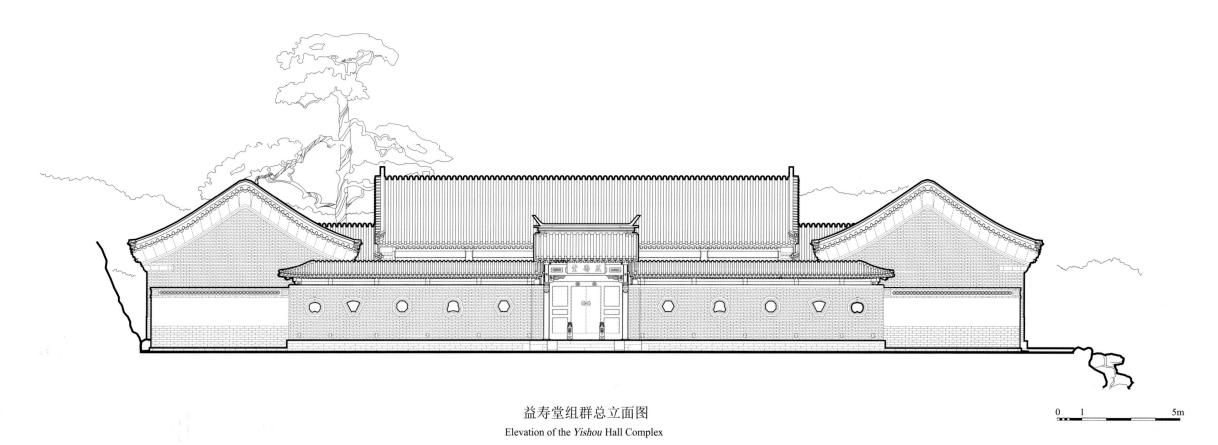

益寿堂组群总立面图

Elevation of the *Yishou* Hall Complex

0 1 5m

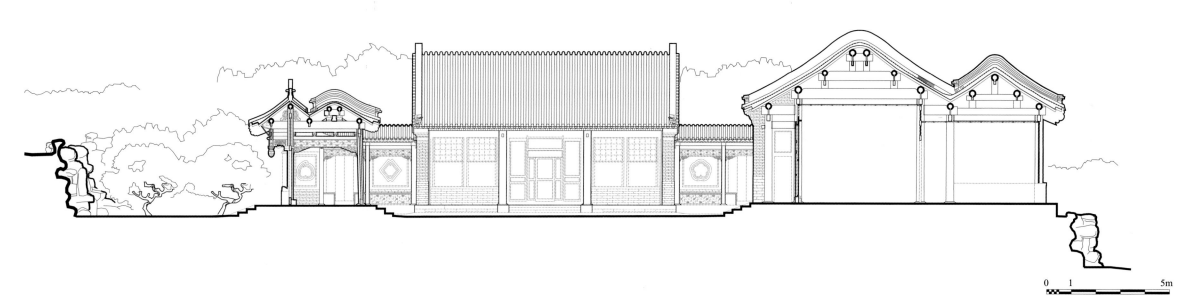

益寿堂组群总剖面图

Section of the *Yishou* Hall Complex

0 1 5m

益寿堂正房正立面图

Front Elevation of the Mail Hall of *Yishou* Hall

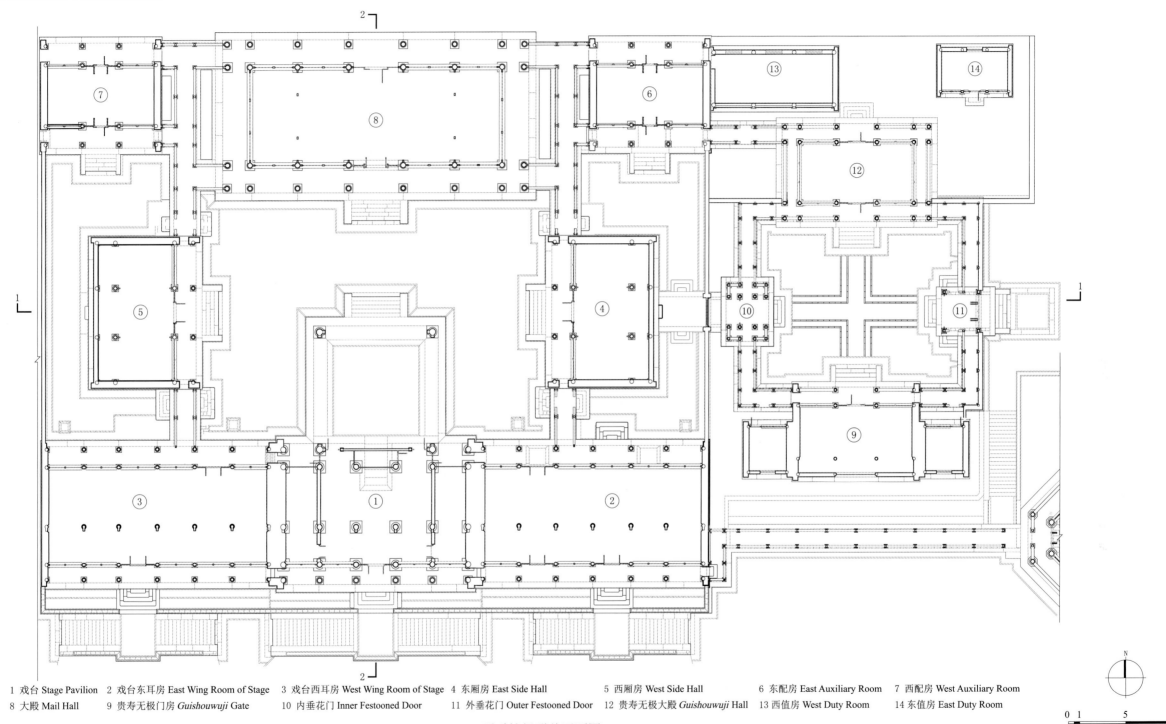

1 戏台 Stage Pavilion　2 戏台东耳房 East Wing Room of Stage　3 戏台西耳房 West Wing Room of Stage　4 东厢房 East Side Hall　5 西厢房 West Side Hall　6 东配房 East Auxiliary Room　7 西配房 West Auxiliary Room

8 大殿 Mail Hall　9 贵寿无极门房 *Guishouwuji* Gate　10 内垂花门 Inner Festooned Door　11 外垂花门 Outer Festooned Door　12 贵寿无极大殿 *Guishouwuji* Hall　13 西值房 West Duty Room　14 东值房 East Duty Room

听鹂馆组群总平面图
Site Plan of the *Tingli* Pavilion Complex

0　1　　　5　　　　　10m

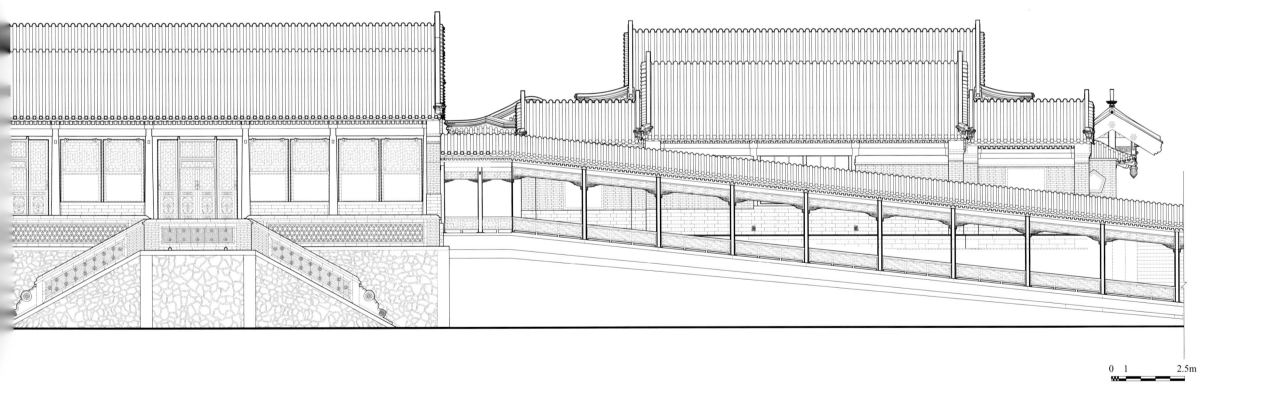

0   1   2.5m

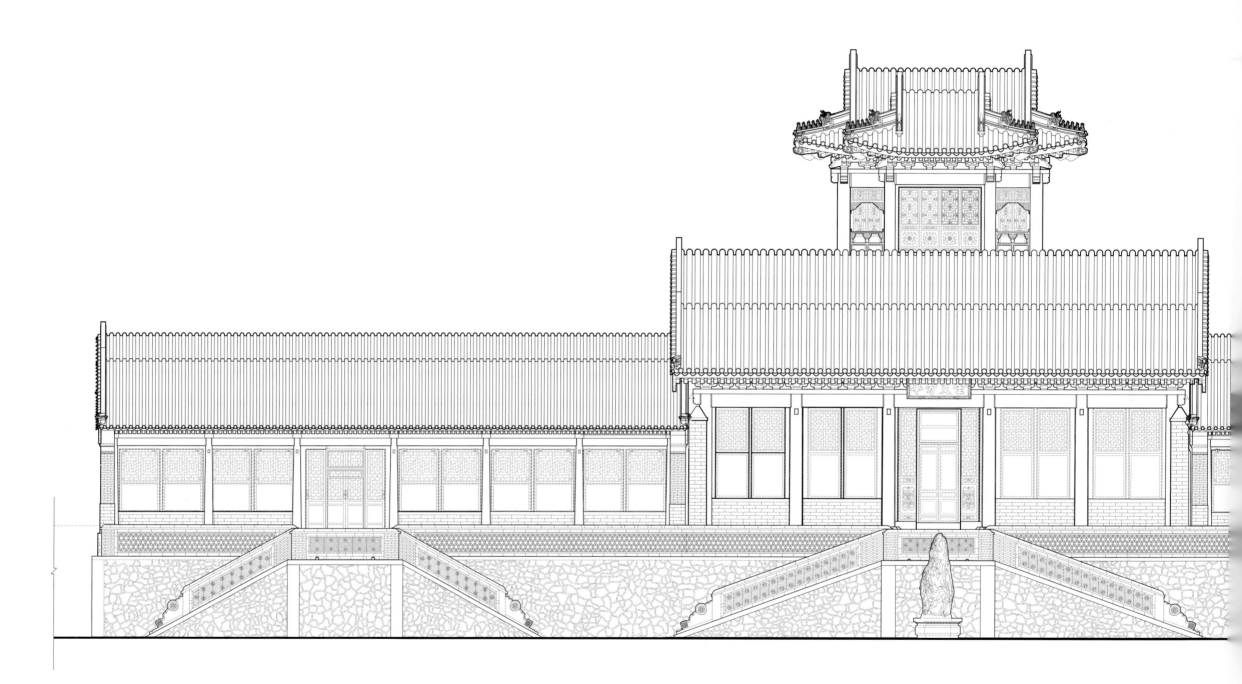

听鹂馆组群总立面图

Elevation of the *Tingli* Pavilion Complex

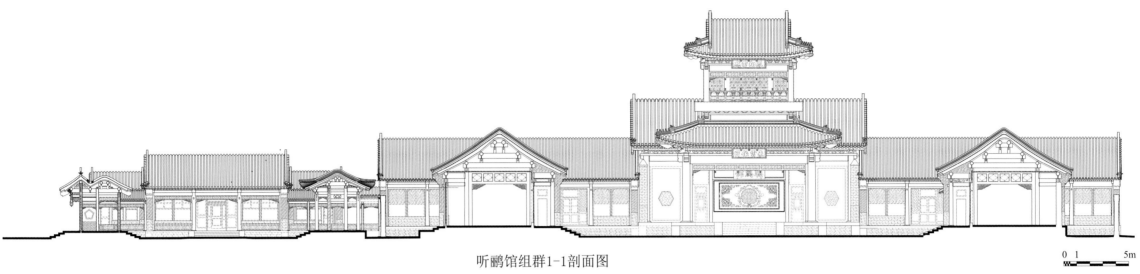

听鹂馆组群1-1剖面图

1-1 Section of the *Tingli* Pavilion Complex

0 1   5m

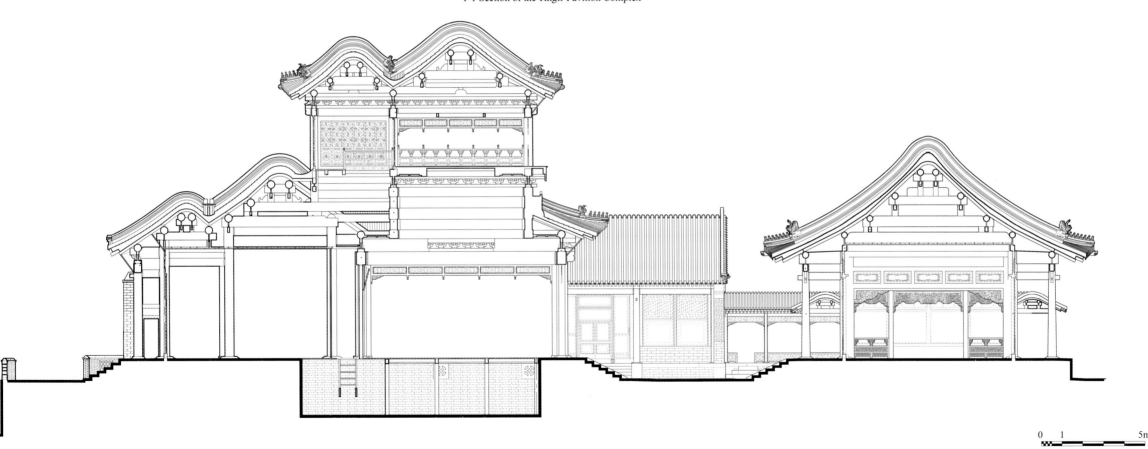

听鹂馆组群2-2剖面图

2-2 Section of the *Tingli* Pavilion Complex

0 1   5m

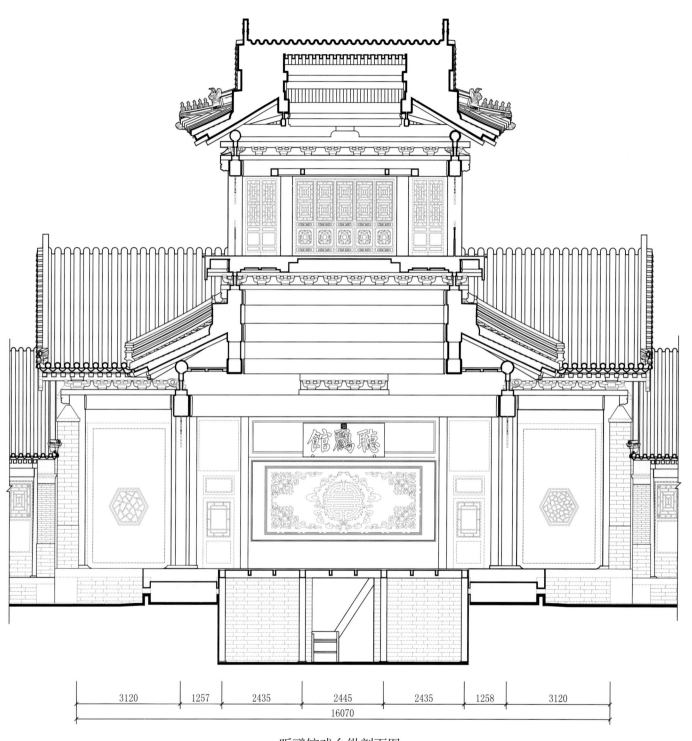

16.000

2745

13.255

1365

11.890

2780

9.110

2515

6.595

1545

5.050

5050

±0.000

1035

−1.035

1690

−2.725

| 3120 | 1257 | 2435 | 2445 | 2435 | 1258 | 3120 |

16070

听鹂馆戏台纵剖面图

Longitudinal Section of the Stage of *Tingli* Pavilion

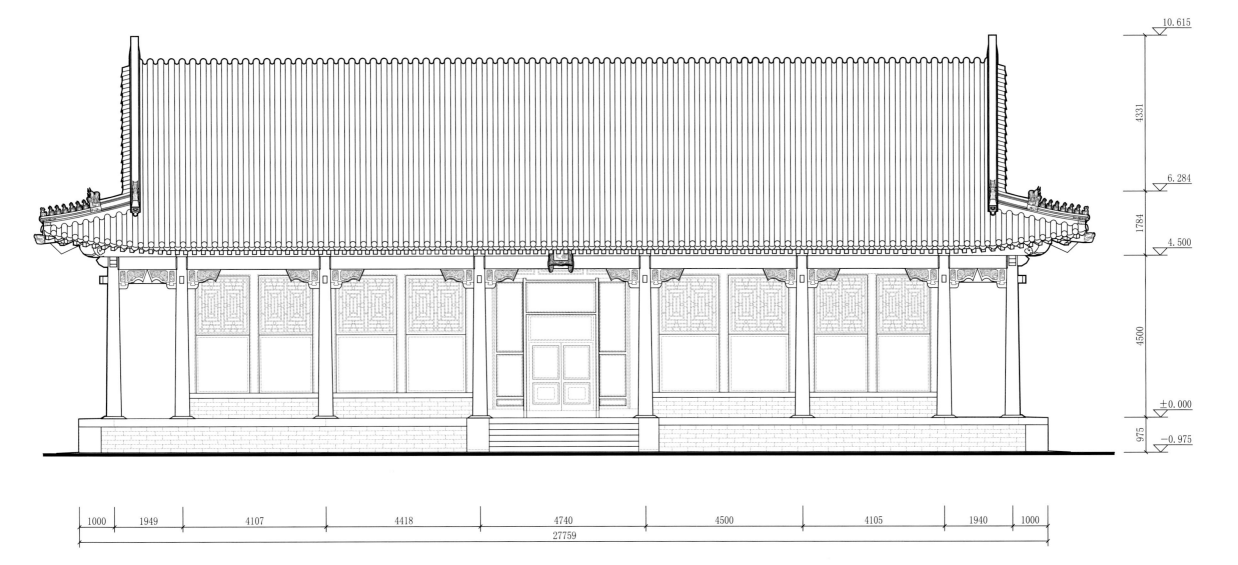

听鹂馆看戏殿正立面图
Front Elevation of the Mail Hall of *Tingli* Pavilion

6.990
280
6.710
3110
3.600
210
3.390
3390
±0.000
625
−0.625

760　1300　3175　3500　3175　1300　760
13970

贵寿无极正立面图
Front Elevation of the *Guishouwuji* Hall

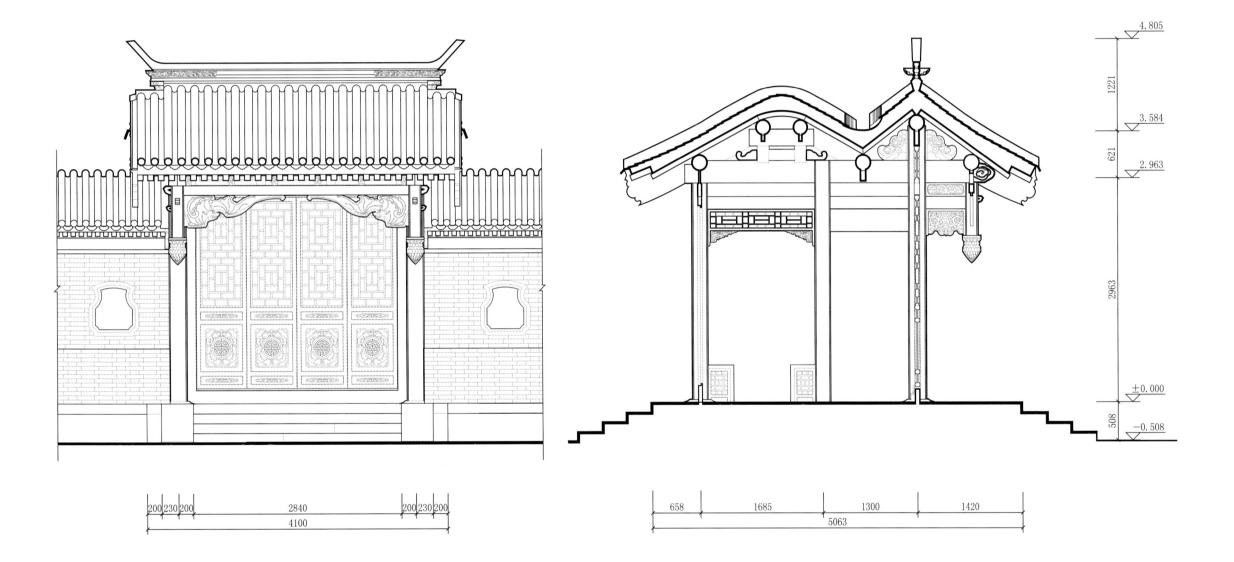

听鹂馆外垂花门正立面图

Front Elevation of the Outer Festooned Door of *Tingli* Pavilion

听鹂馆外垂花门剖面图

Section of the Outer Festooned Door of *Tingli* Pavilion

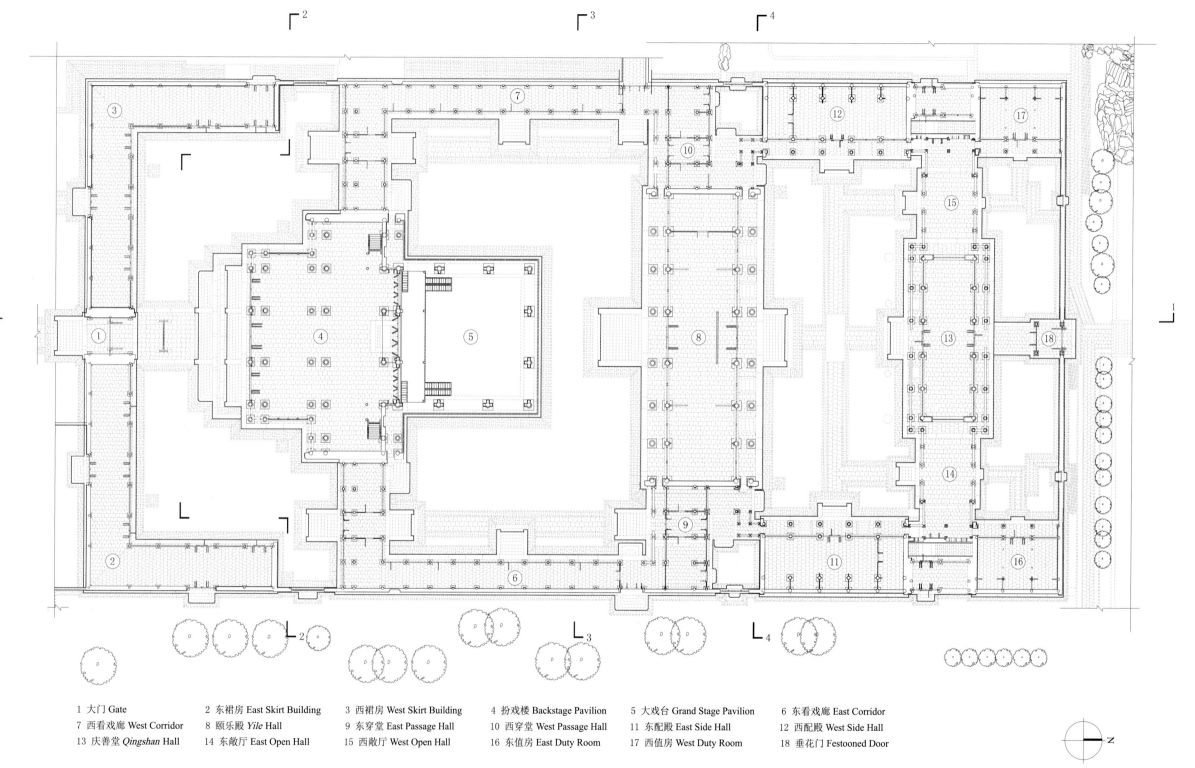

1 大门 Gate   2 东裙房 East Skirt Building   3 西裙房 West Skirt Building   4 扮戏楼 Backstage Pavilion   5 大戏台 Grand Stage Pavilion   6 东看戏廊 East Corridor

7 西看戏廊 West Corridor   8 颐乐殿 *Yile* Hall   9 东穿堂 East Passage Hall   10 西穿堂 West Passage Hall   11 东配殿 East Side Hall   12 西配殿 West Side Hall

13 庆善堂 *Qingshan* Hall   14 东敞厅 East Open Hall   15 西敞厅 West Open Hall   16 东值房 East Duty Room   17 西值房 West Duty Room   18 垂花门 Festooned Door

德和园组群总平面图
Site Plan of the *Dehe* Garden Complex

0 1   5   10m

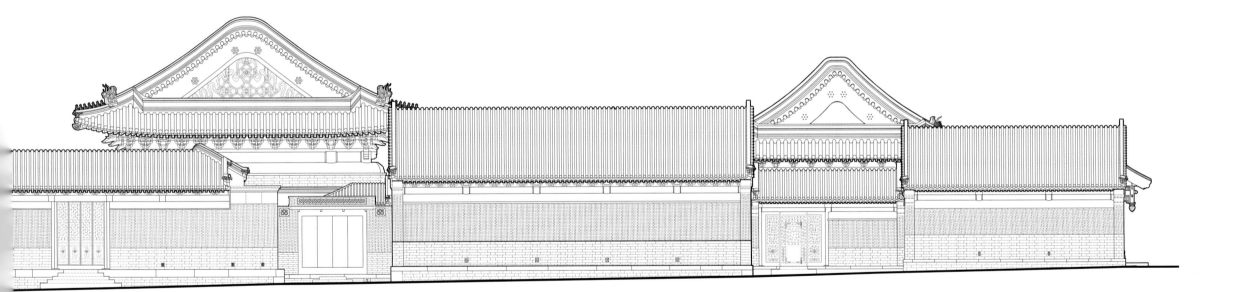

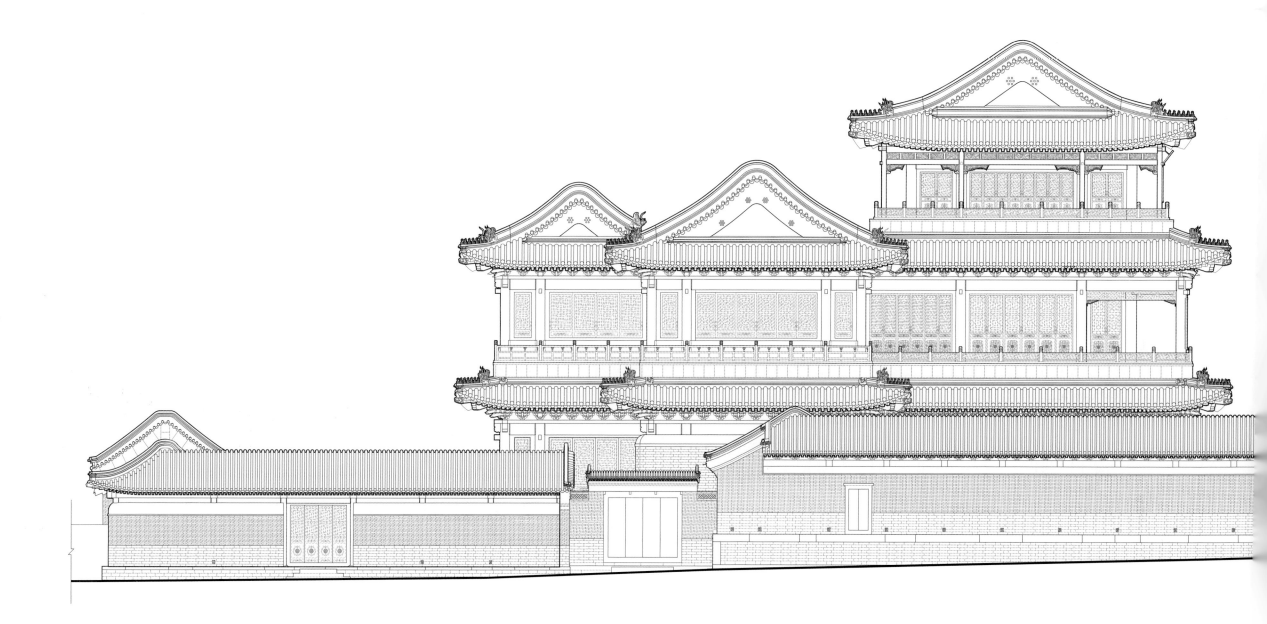

德和园组群侧立面图

Side Elevation of the *Dehe* Garden Complex

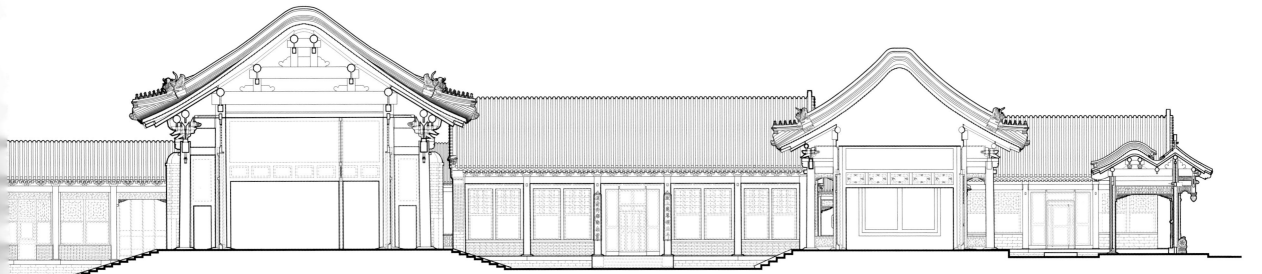

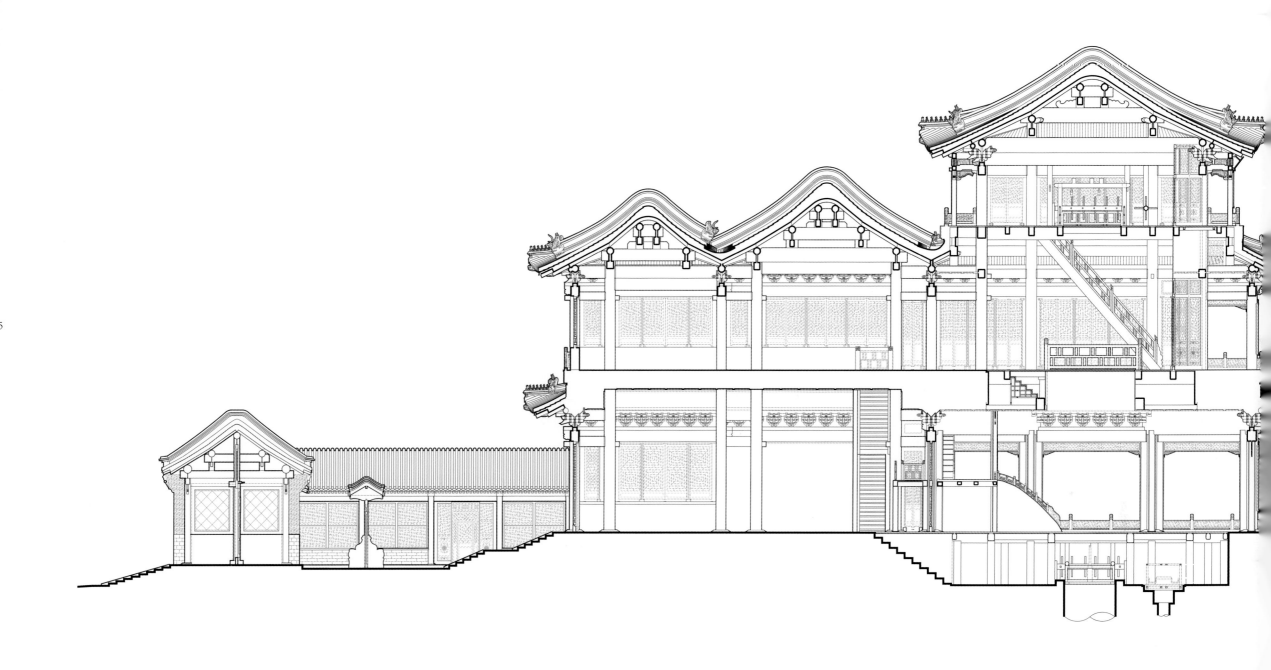

德和园组群1-1剖面图

1-1 Section of the *Dehe* Garden Complex

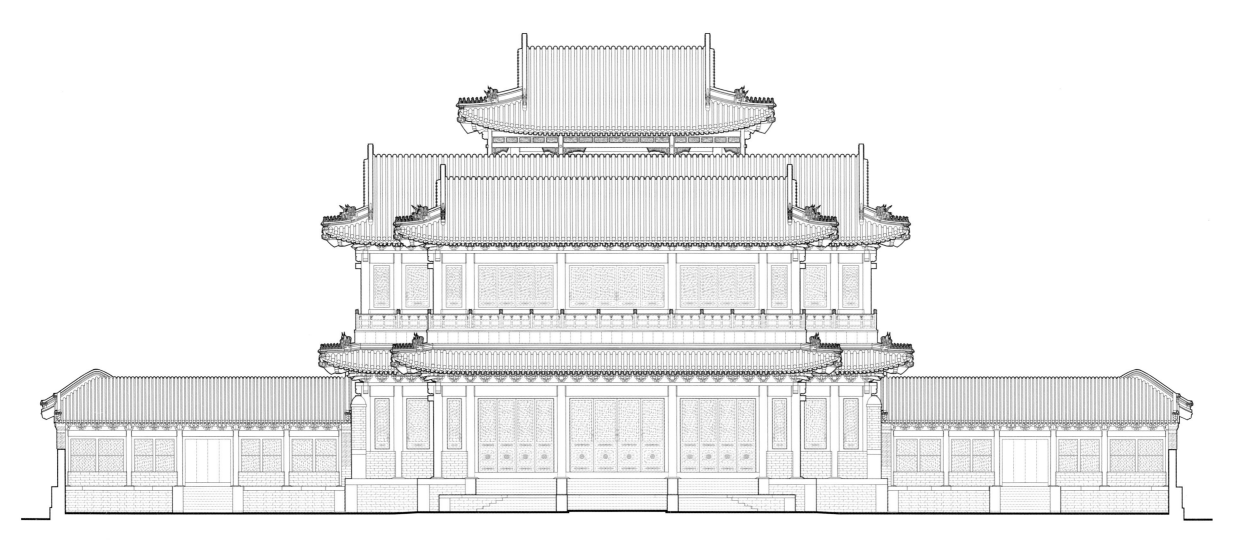

德和园组群2-2剖面图

2-2 Section of the *Dehe* Garden Complex

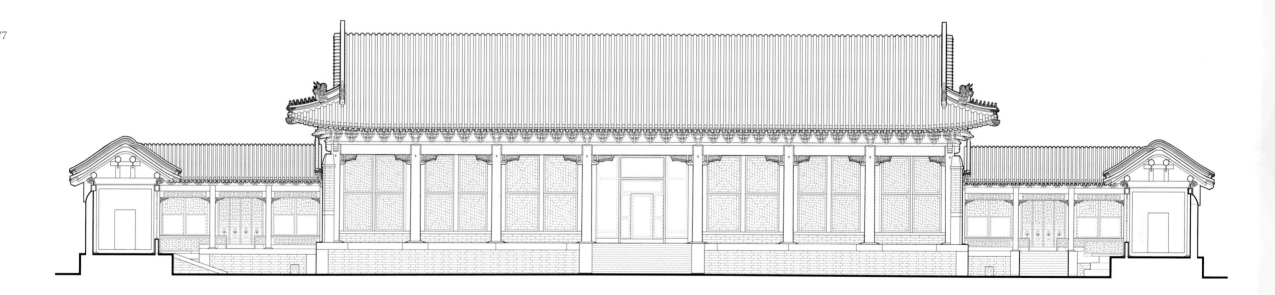

德和园组群3-3剖面图

3-3 Section of the *Dehe* Garden Complex

0 1 5m

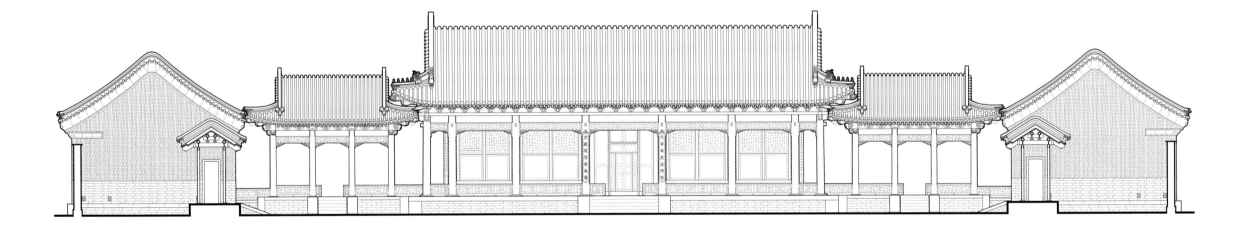

德和园组群4-4剖面图

4-4 Section of the *Dehe* Garden Complex

0　1　　　　5m

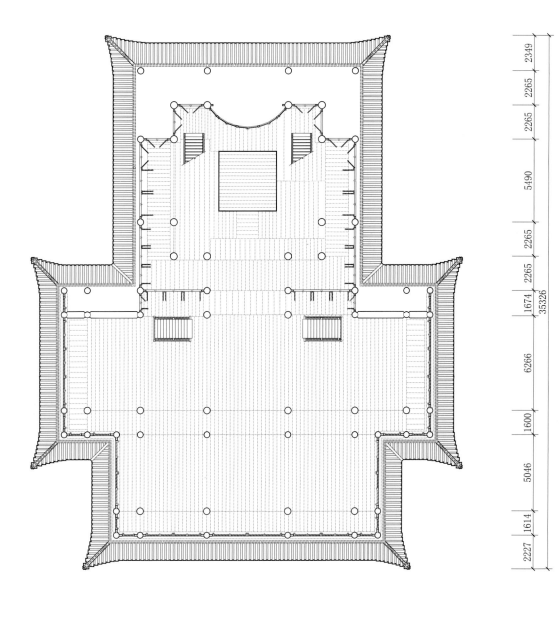

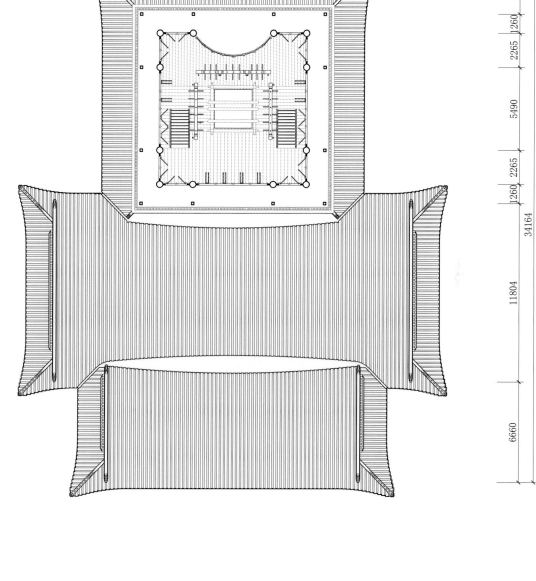

大戏楼二层平面图
Second Floor Plan of the Grand Stage Pavilion

大戏楼三层平面图
Third Floor Plan of the Grand Stage Pavilion

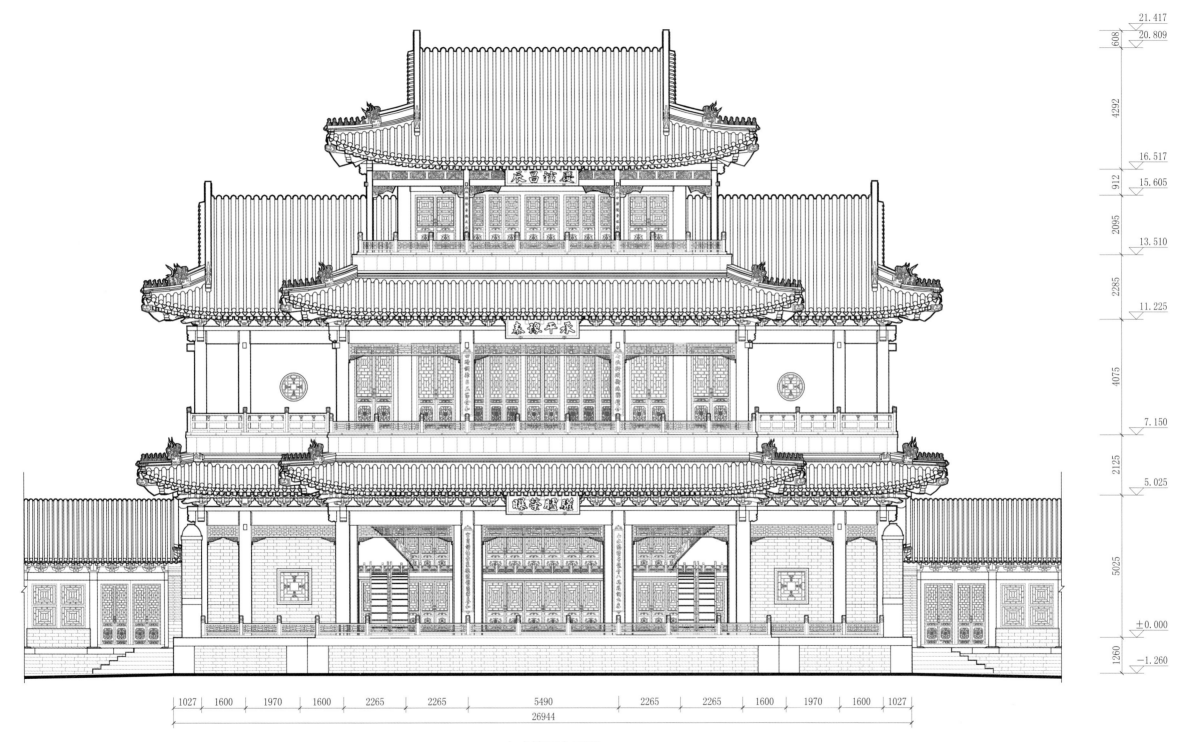

21.417
20.809
608
4292
16.517
15.605
912
2095
13.510
2285
11.225
4075
7.150
2125
5.025
5025
±0.000
1260
−1.260

| 1027 | 1600 | 1970 | 1600 | 2265 | 2265 | 5490 | 2265 | 2265 | 1600 | 1970 | 1600 | 1027 |

26944

大戏楼正立面图
Front Elevation of the Grand Stage Pavilion

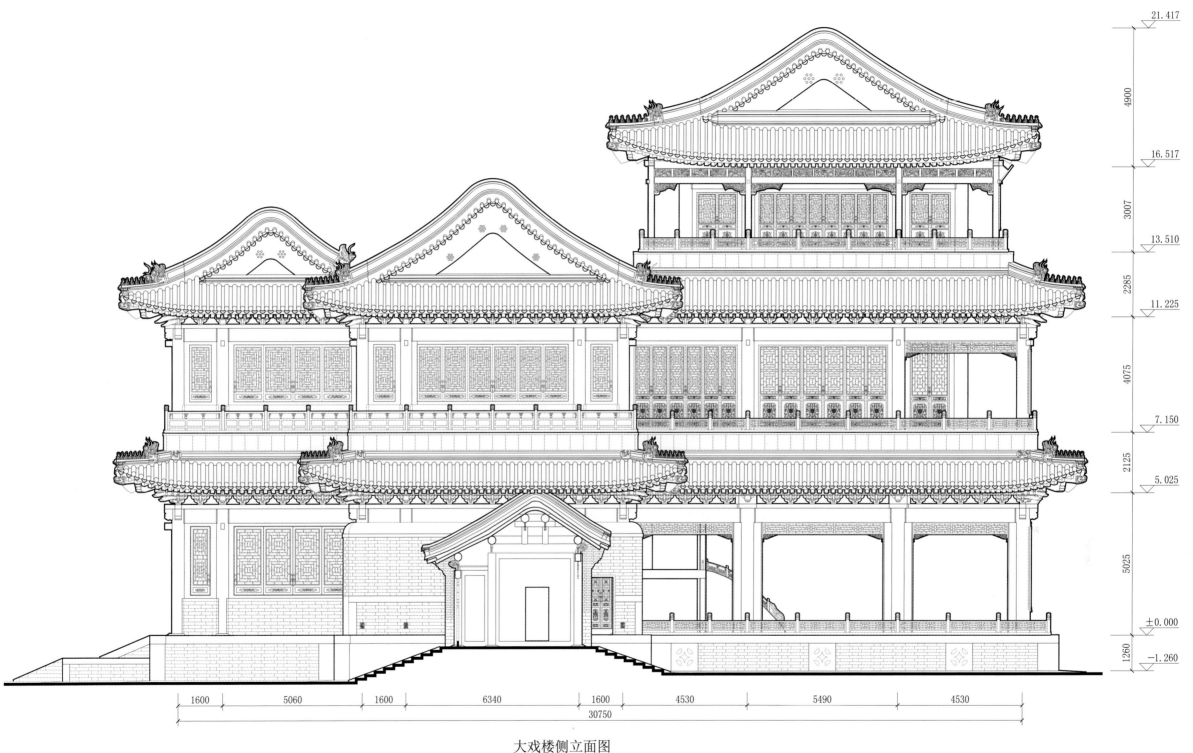

大戏楼侧立面图

Side Elevation of the Grand Stage Pavilion

0 1 5m

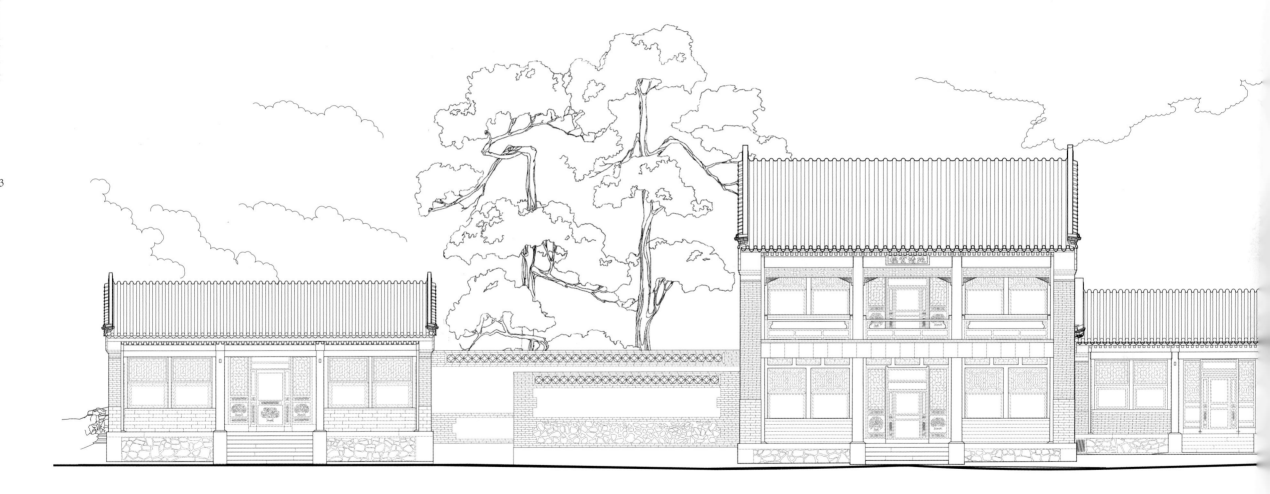

西所买卖街东侧建筑群立面图

Elevation of the East Complex of West Shopping Street

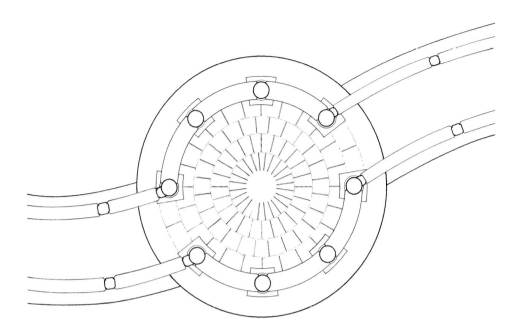

小有天平面图（1957 年测绘图）
Plan of the *Xiaoyoutian*（Drawn in 1957）

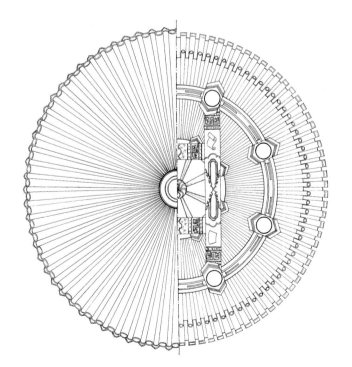

小有天梁架仰俯视图（1957 年测绘图）
Framework Plan of the *Xiaoyoutian*（Drawn in 1957）

0   1   2m

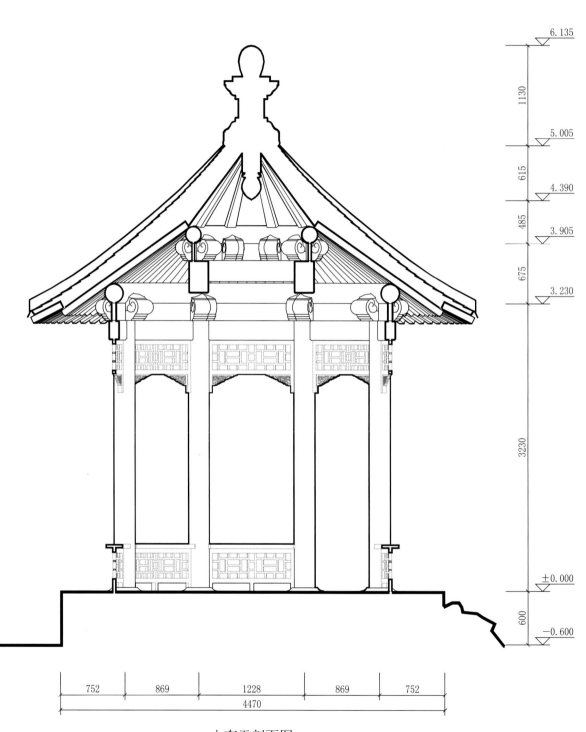

小有天剖面图
Section of the *Xiaoyoutian*

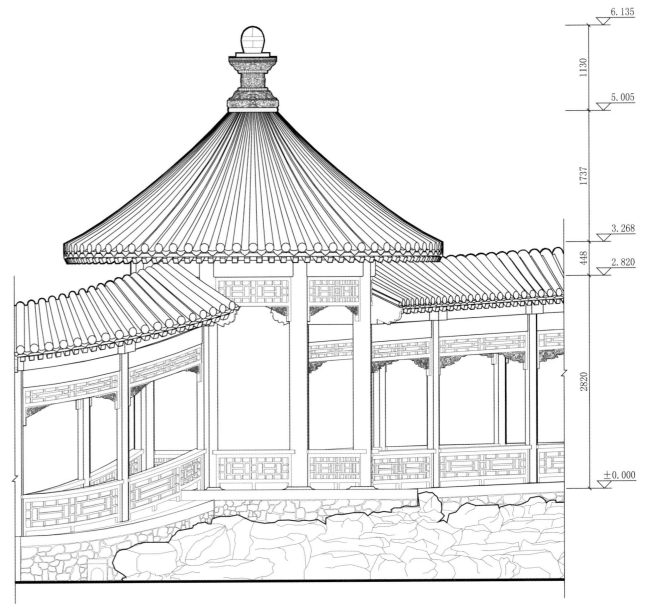

6.135

1130

5.005

1737

3.268

448

2.820

2820

±0.000

小有天立面图（1957 年测绘图）

Elevation of the *Xiaoyoutian* (Drawn in 1957)

0   0.5   1m

869   1228   869

2966

小有天正立面图

Front Elevation of the *Xiaoyoutian*

延清赏楼正立面图
Front Elevation of the *Yanqingshang* Pavilion

延清赏楼明间剖面图
Mingjian Section of the *Yanqingshang* Pavilion

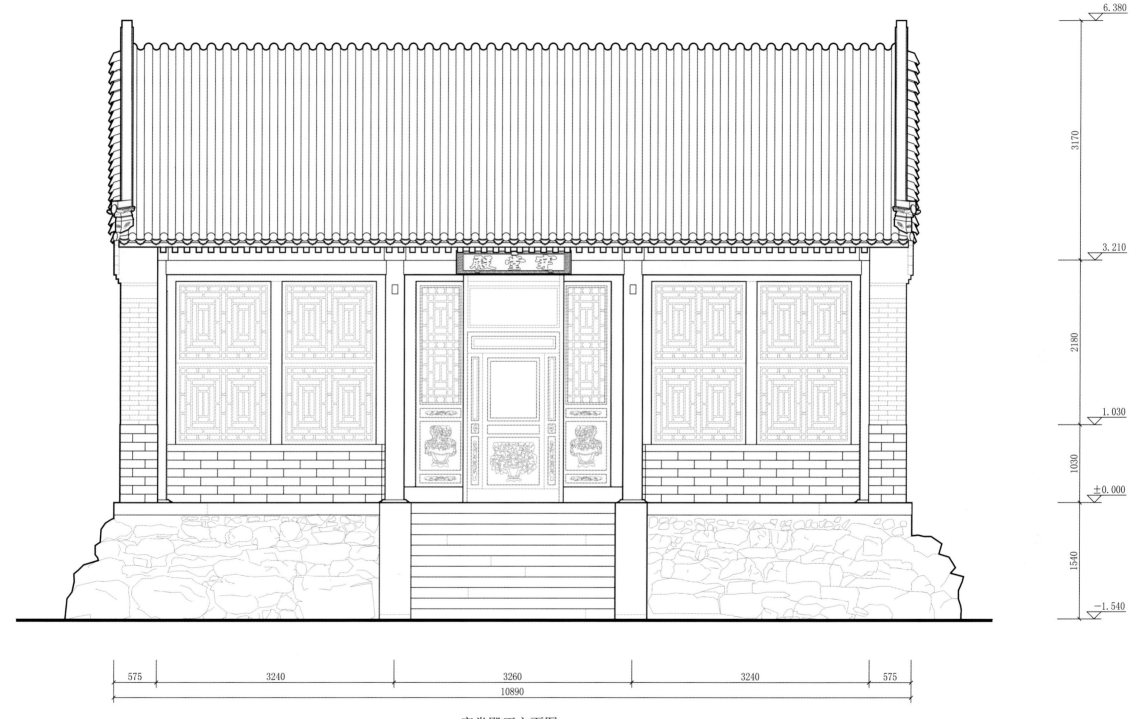

穿堂殿正立面图
Front Elevation of the Passage Hall

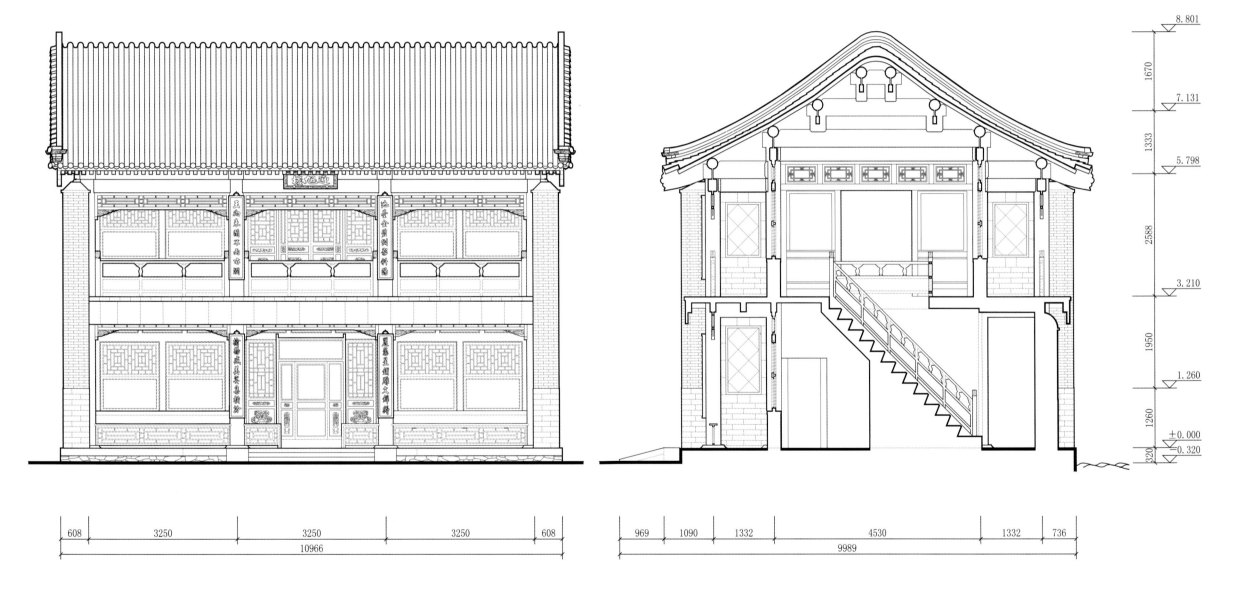

迎旭楼正立面图

Front Elevation of the *Yingxu* Pavilion

迎旭楼剖面图

Section of the *Yingxu* Pavilion

乡村建筑
Rural Architecture

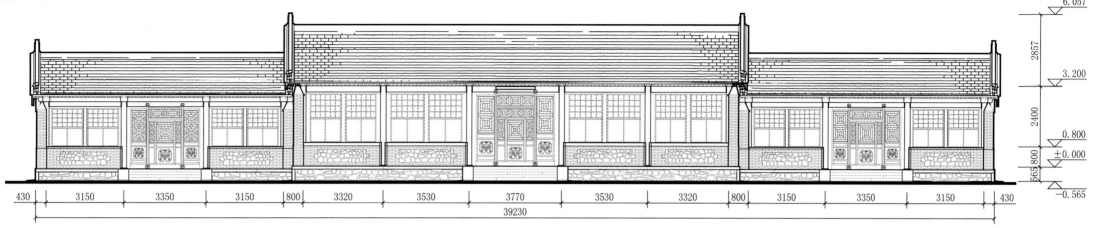

乐农轩正立面图
Front Elevation of the *Lenong* Pavilion

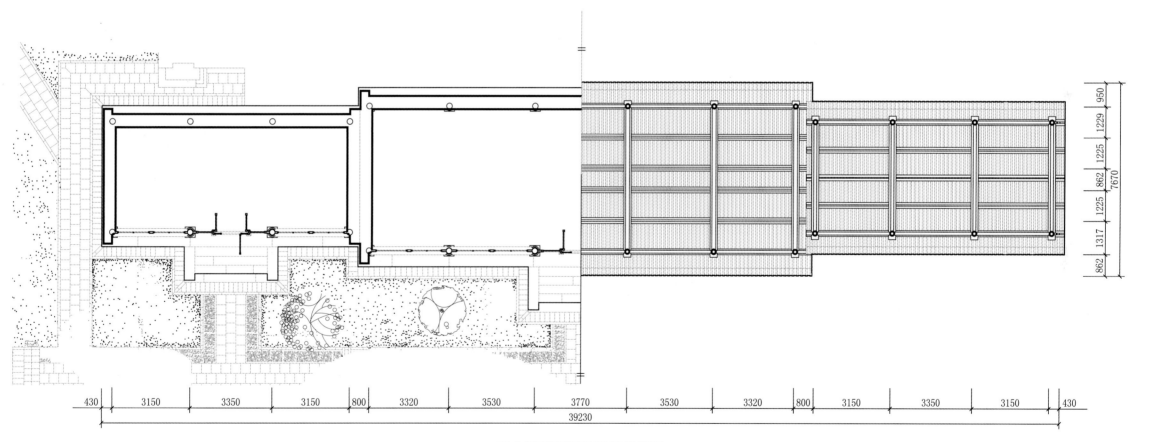

乐农轩平面图及梁架仰视图
Plan and Framework Plan of the *Lenong* Pavilion

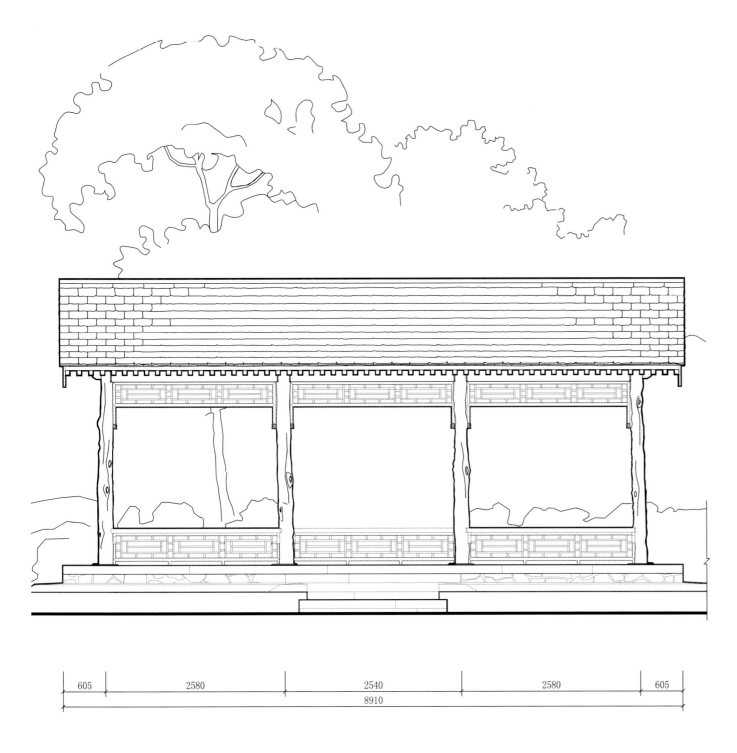

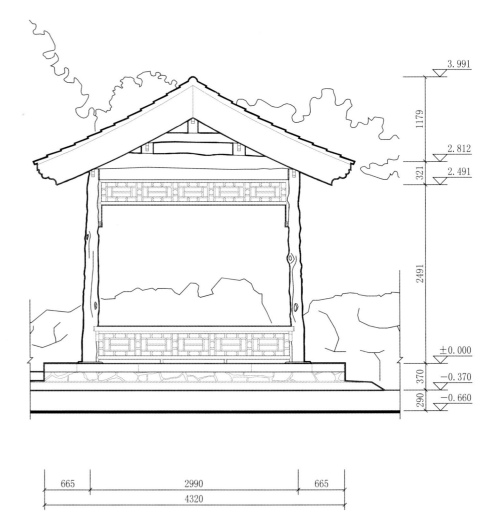

3.991

1179

2.812

321

2.491

190

2491

±0.000

370

−0.370

290

−0.660

605　2580　2540　2580　605

8910

665　2990　665

4320

草亭正立面图

Front Elevation of the *Cao* Pavilion

草亭侧立面图

Side Elevation of the *Cao* Pavilion

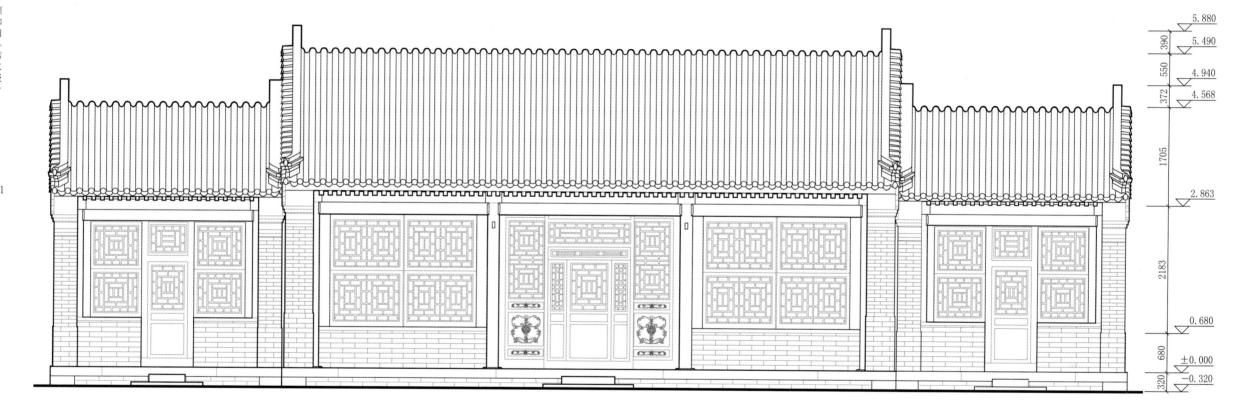

自在庄正立面图

Front Elevation of the *Zizai* Hall

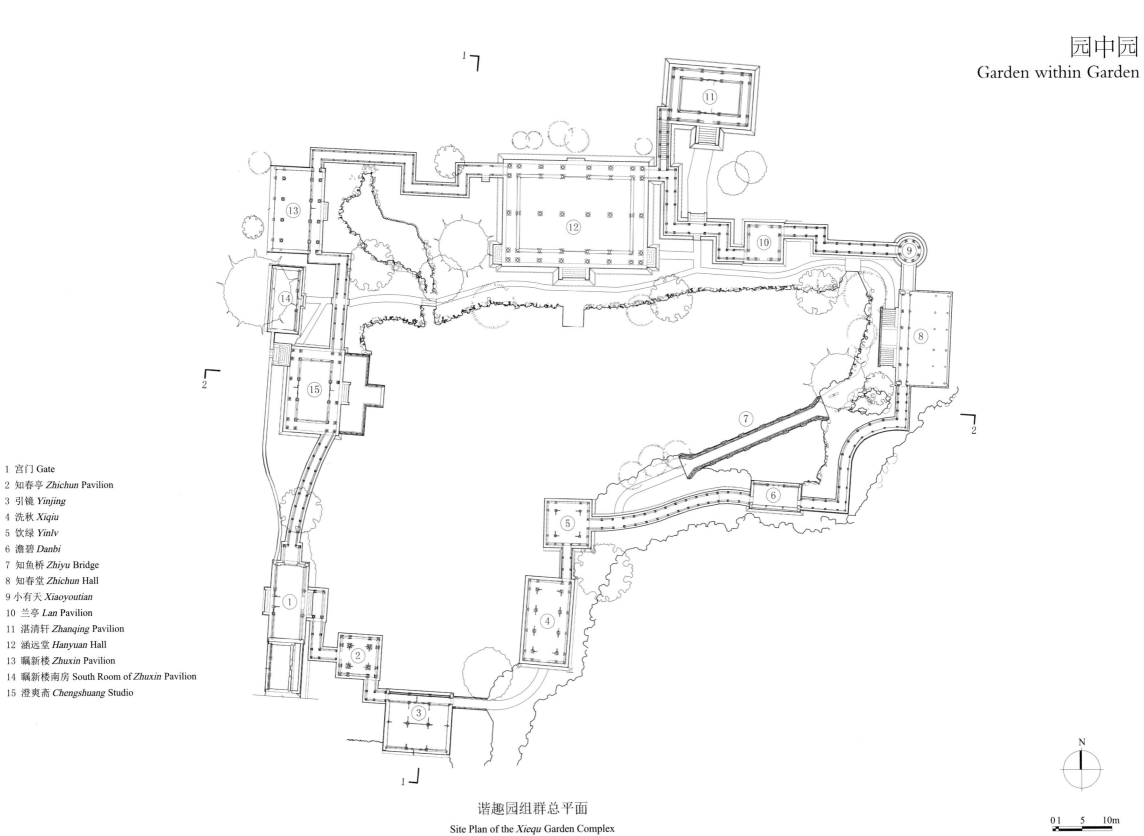

1　宫门 Gate
2　知春亭 *Zhichun* Pavilion
3　引镜 *Yinjing*
4　洗秋 *Xiqiu*
5　饮绿 *Yinlv*
6　潋碧 *Danbi*
7　知鱼桥 *Zhiyu* Bridge
8　知春堂 *Zhichun* Hall
9　小有天 *Xiaoyoutian*
10　兰亭 *Lan* Pavilion
11　湛清轩 *Zhanqing* Pavilion
12　涵远堂 *Hanyuan* Hall
13　瞩新楼 *Zhuxin* Pavilion
14　瞩新楼南房 South Room of *Zhuxin* Pavilion
15　澄爽斋 *Chengshuang* Studio

谐趣园组群总平面

Site Plan of the *Xiequ* Garden Complex

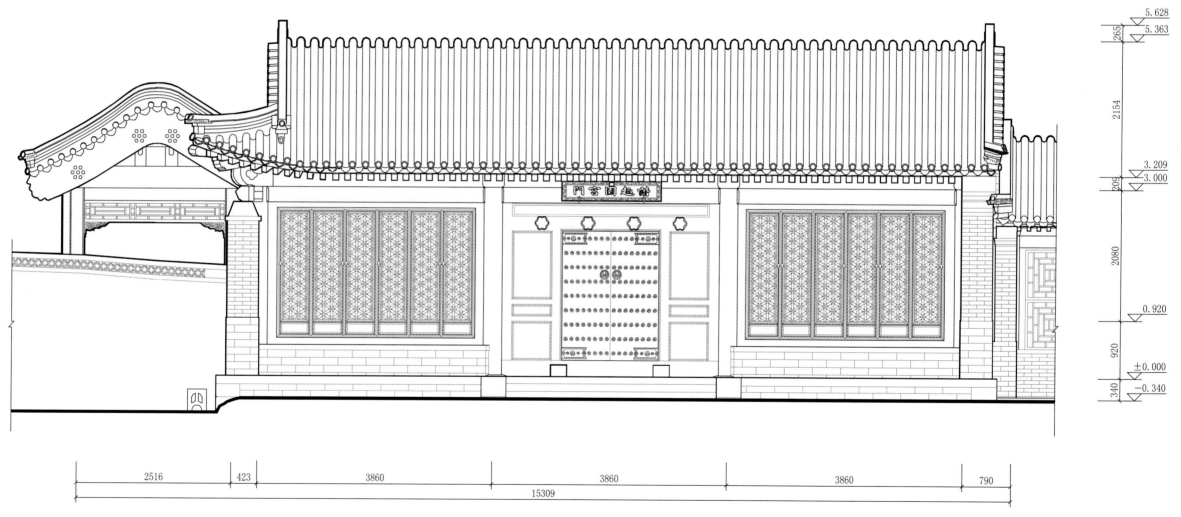

谐趣园宫门正立面

Front Elevation of the Gate of *Xiequ* Garden

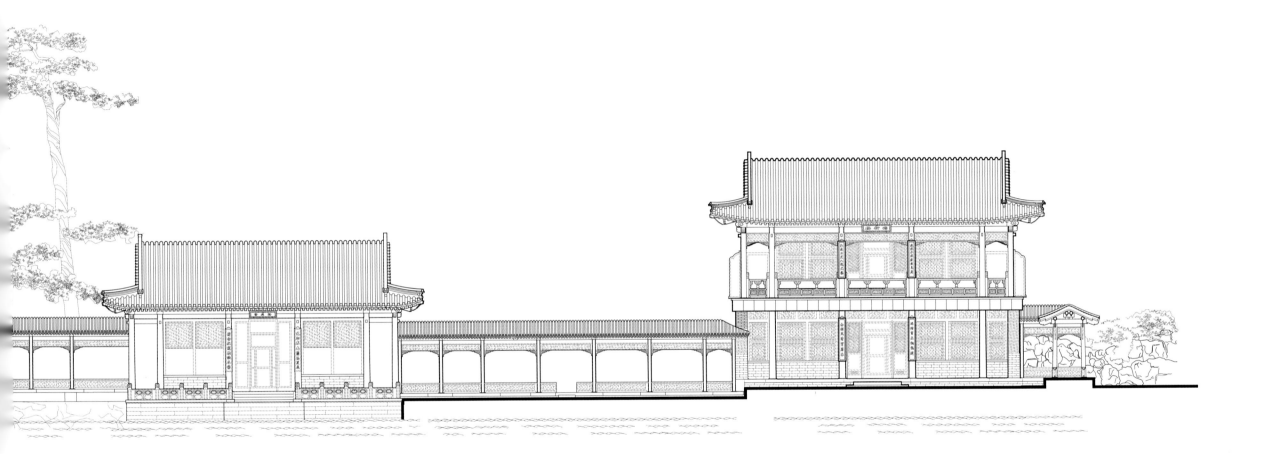

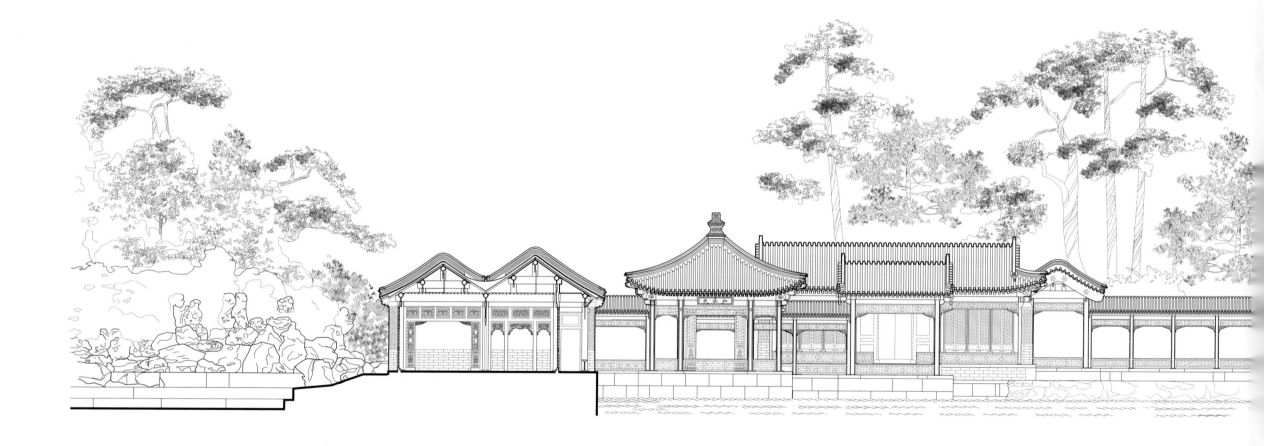

谐趣园组群1-1剖面图

1-1 Section of the *Xiequ* Garden Complex

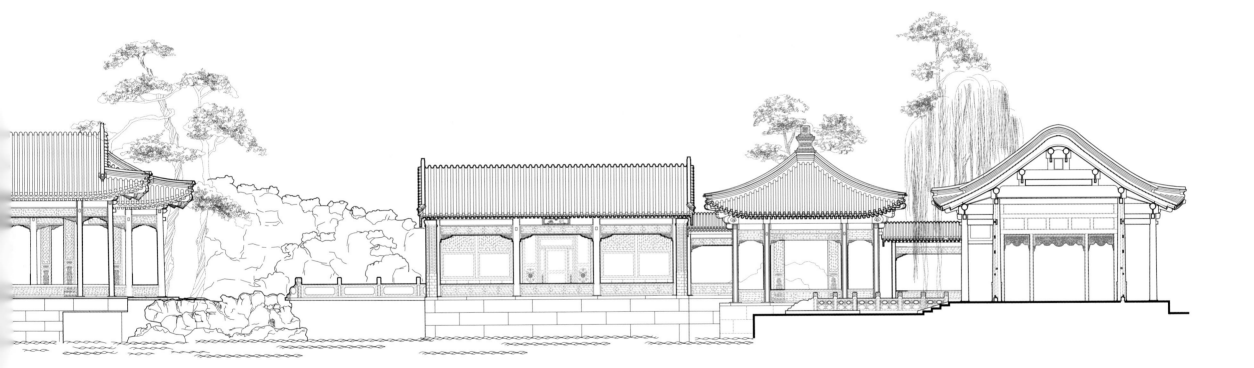

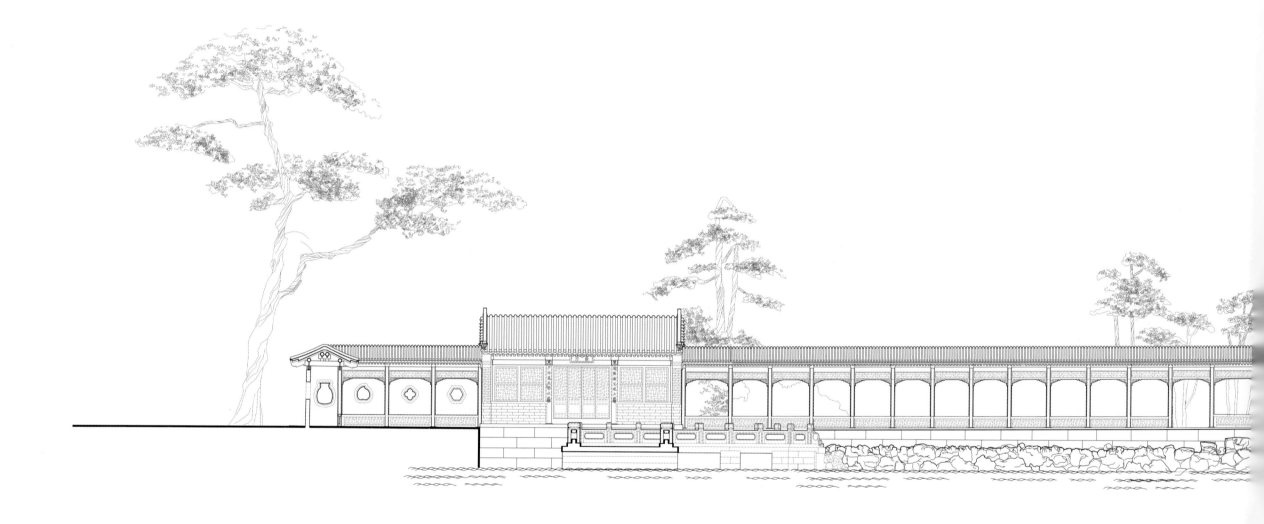

谐趣园组群2-2剖面图

2-2 Section of the *Xiequ* Garden Complex

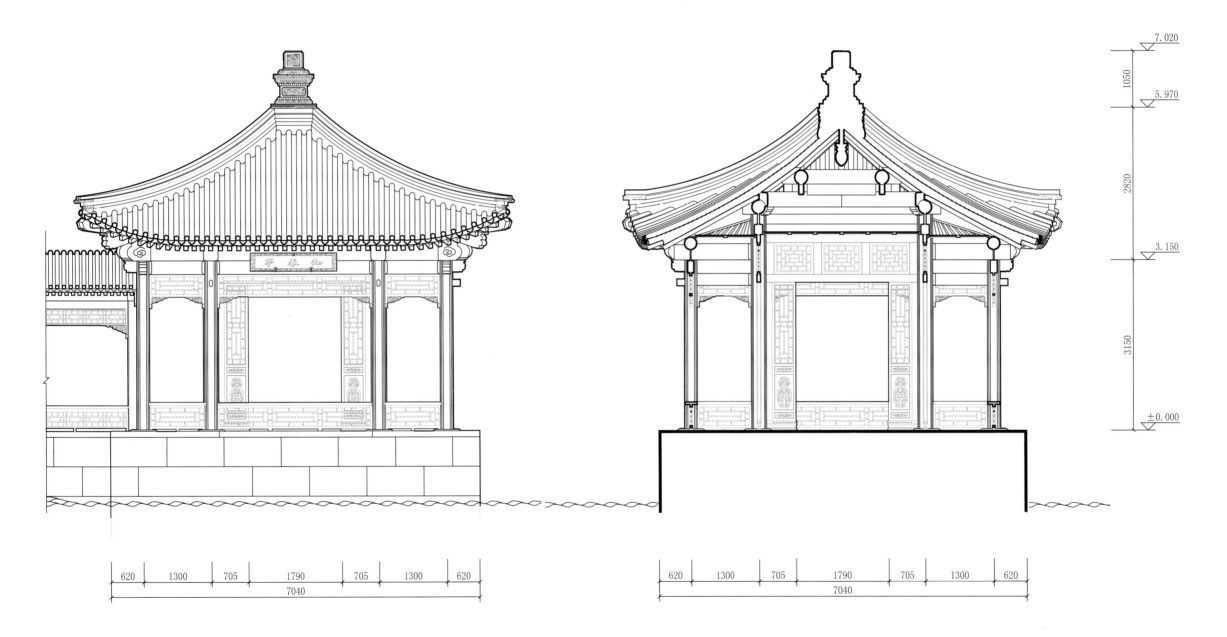

知春亭正立面图

Front Elevation of the *Zhichun* Pavilion

知春亭剖面图

Section of the *Zhichun* Pavilion

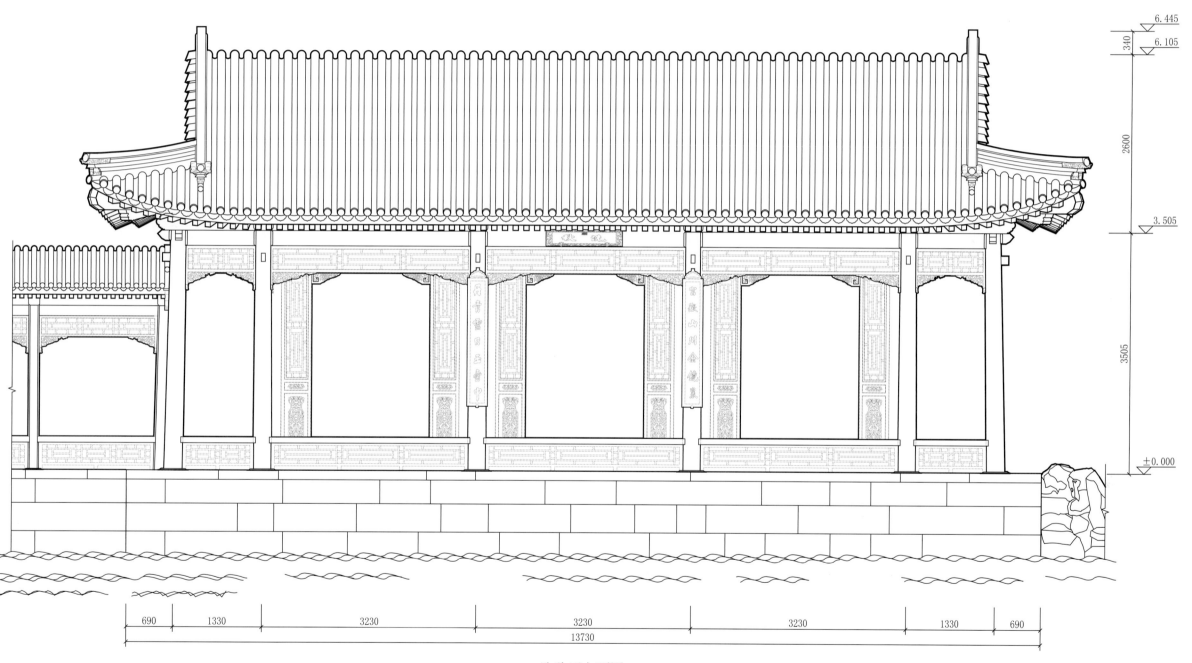

洗秋正立面图

Front Elevation of *Xiqiu*

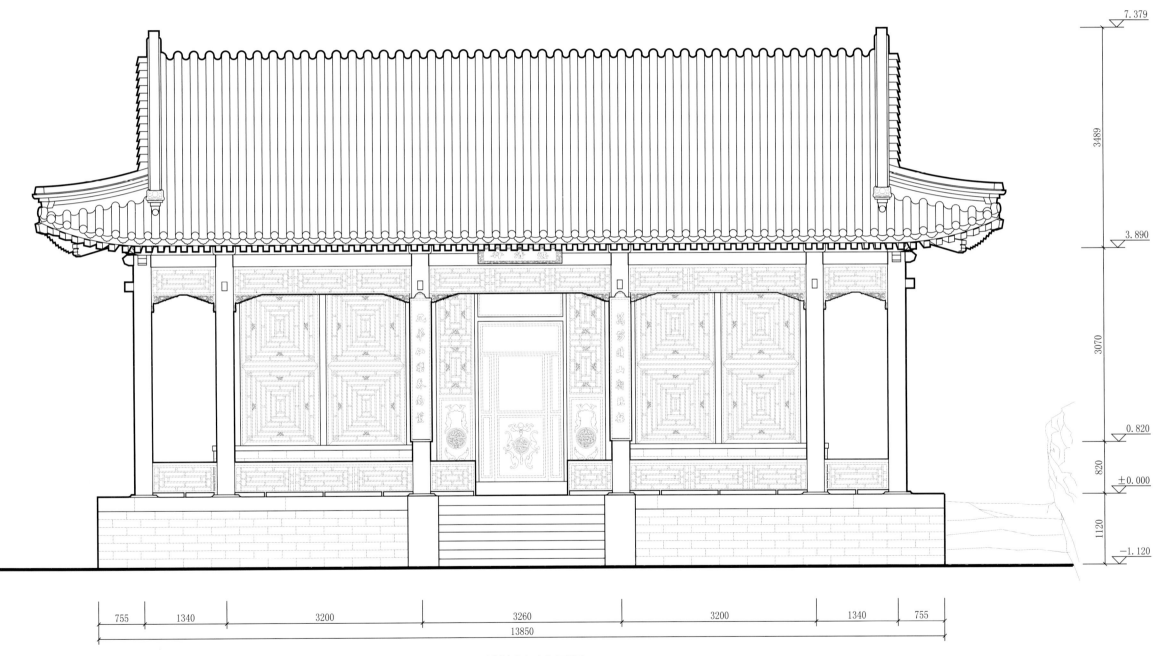

7.379

3489

3.890

3070

0.820

820

±0.000

1120

−1.120

755 1340 3200 3260 3200 1340 755

13850

湛清轩正立面图
Front Elevation of *Zhanqing* Pavilion

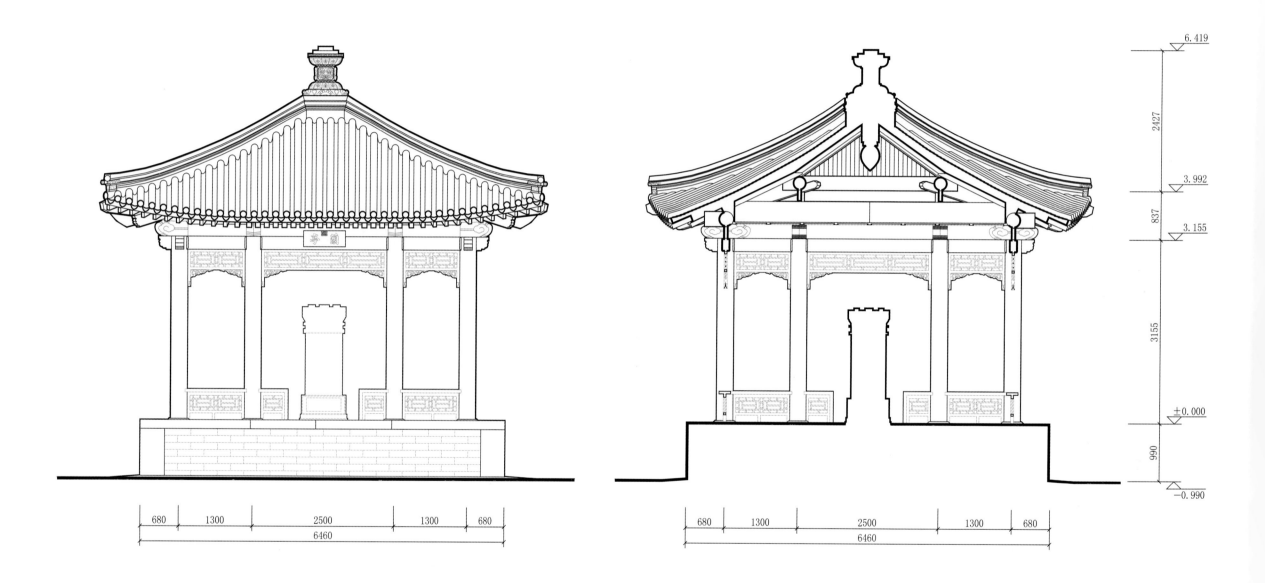

6.419

2427

3.992

837

3.155

3155

±0.000

990

−0.990

| 680 | 1300 | 2500 | 1300 | 680 |

6460

| 680 | 1300 | 2500 | 1300 | 680 |

6460

兰亭正立面图

Front Elevation of the *Lan* Pavilion

兰亭剖面图

Section of the *Lan* Pavilion

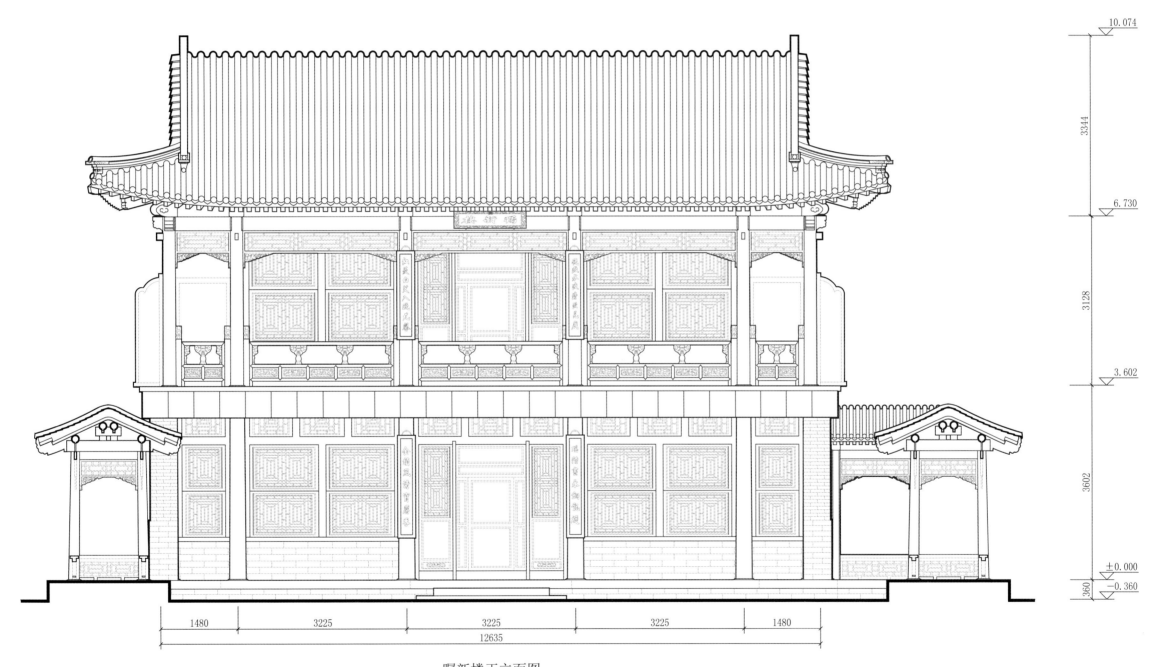

10.074

3344

6.730

3128

3.602

3602

±0.000

360

−0.360

1480 3225 3225 3225 1480

12635

瞩新楼正立面图

Front Elevation of the *Zhuxin* Pavilion

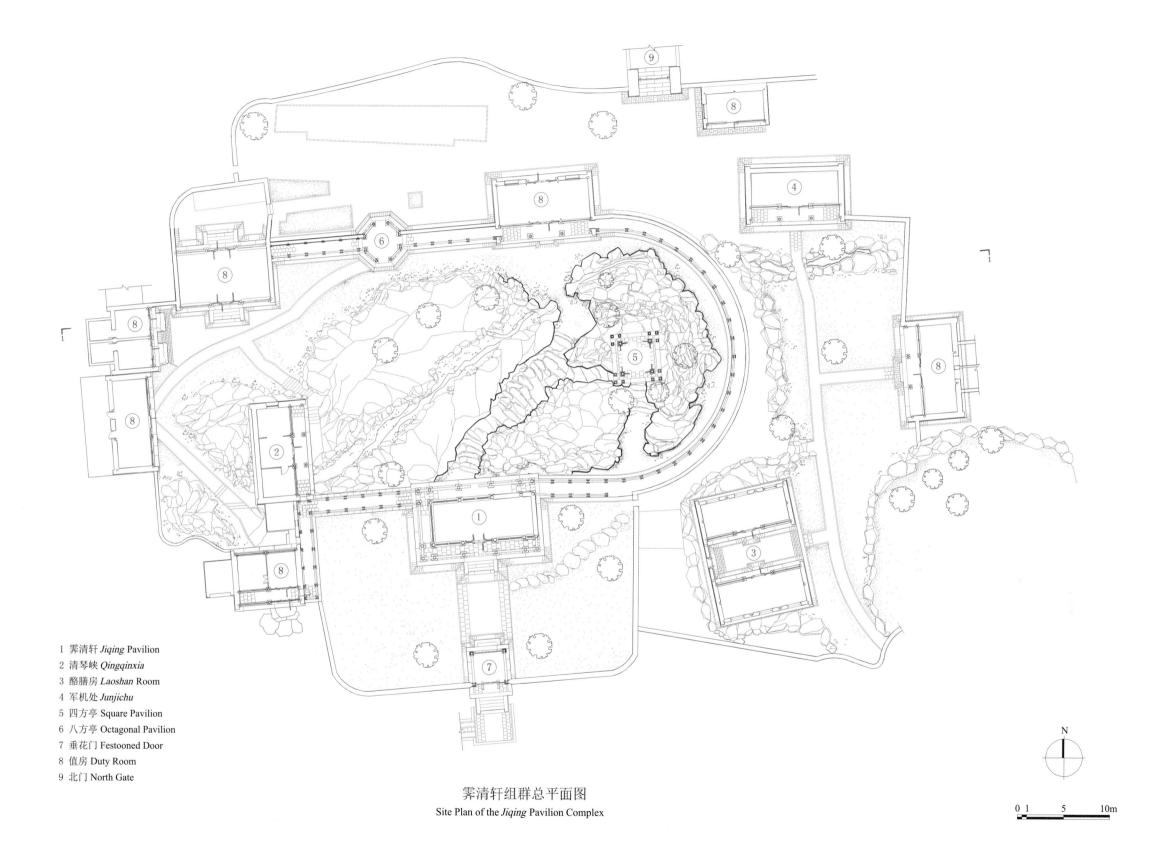

1　霁清轩 *Jiqing* Pavilion
2　清琴峡 *Qingqinxia*
3　酪膳房 *Laoshan* Room
4　军机处 *Junjichu*
5　四方亭 Square Pavilion
6　八方亭 Octagonal Pavilion
7　垂花门 Festooned Door
8　值房 Duty Room
9　北门 North Gate

霁清轩组群总平面图
Site Plan of the *Jiqing* Pavilion Complex

N

0 1　　5　　10m

0　1　　　　　5m

霁清轩组群1-1剖面图

1-1 Section of the *Jiqing* Pavilion Complex

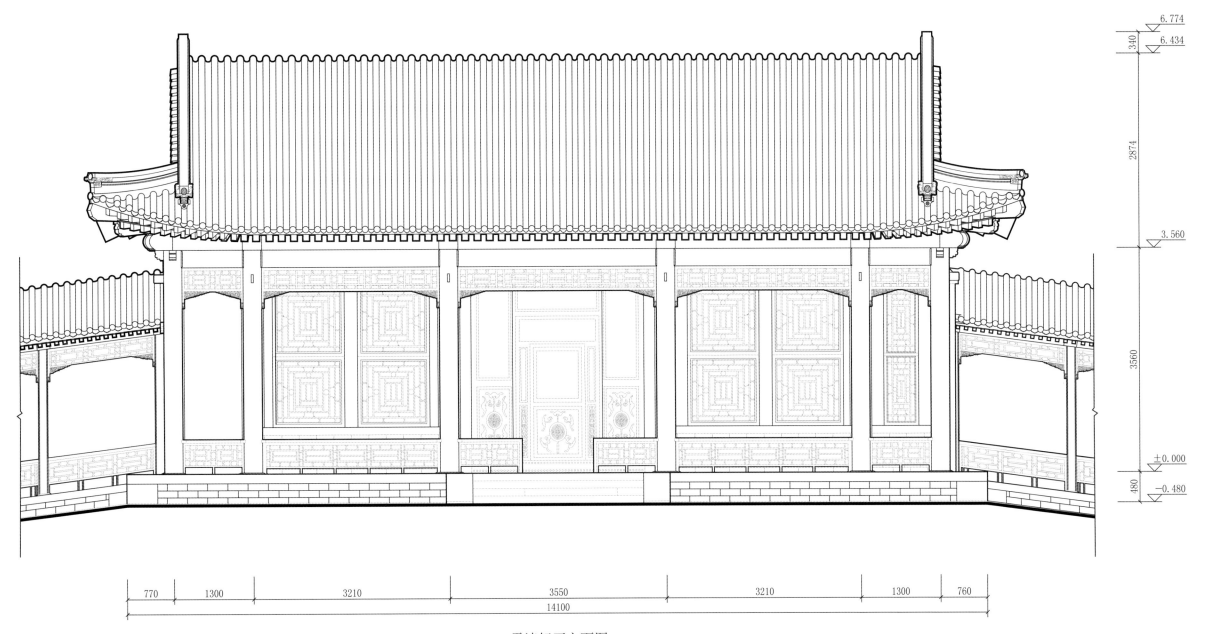

6.774

340

6.434

2874

3.560

3560

±0.000

480

−0.480

| 770 | 1300 | 3210 | 3550 | 3210 | 1300 | 760 |

14100

霁清轩正立面图

Front Elevation of the *Jiqing* Pavilion

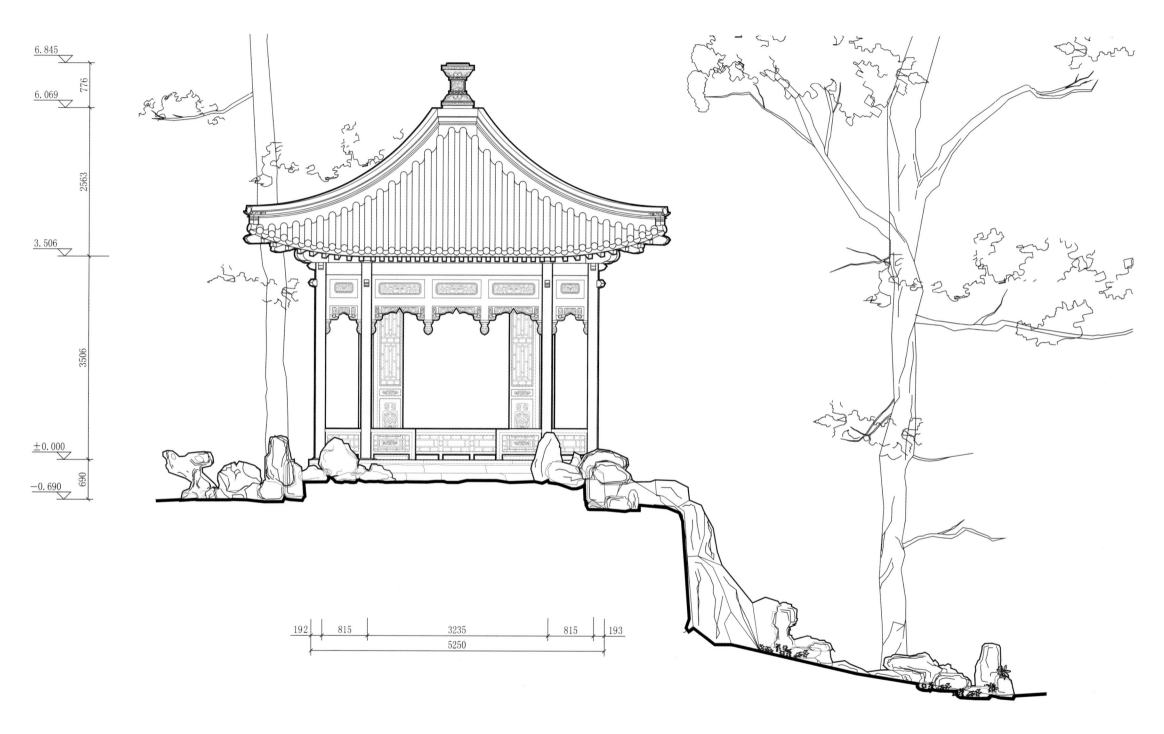

6.845

6.069

776

2563

3.506

3506

±0.000

690

−0.690

192 815 3235 815 193

5250

四方亭侧立面图

Side Elevation of the Square Pavilion

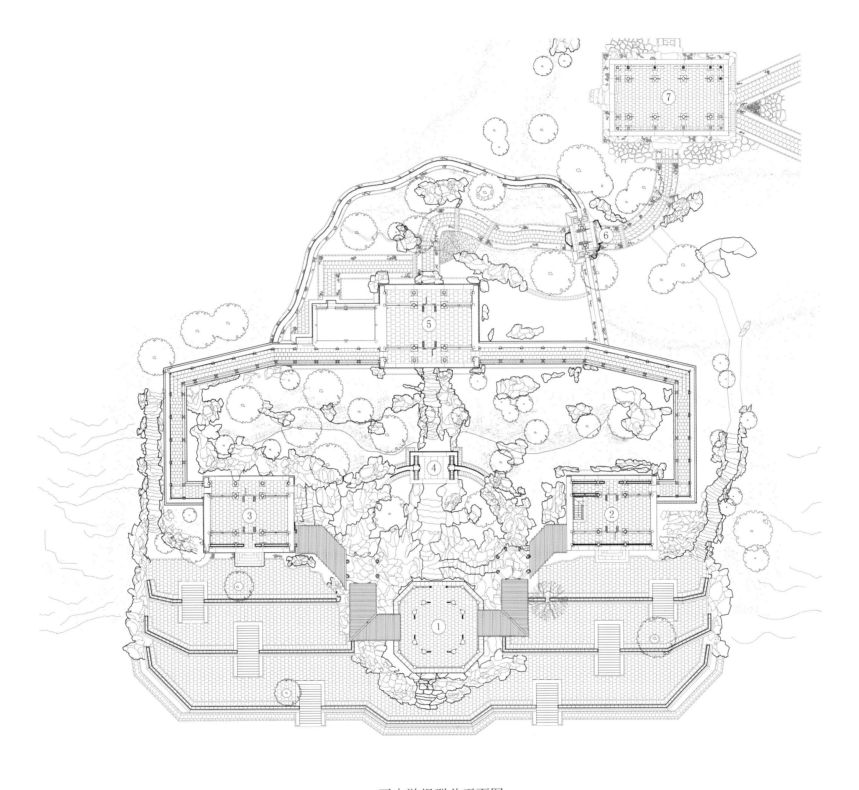

1 澄辉阁 *Chenghui* Pavilion
2 爱山楼 *Aishan* Pavilion
3 借秋楼 *Jieqiu* Pavilion
4 牌楼 Pailou
5 画中游 *Huazhongyou*
6 垂花门 Festooned Door
7 湖山真意 *Hushanzhenyi*

N

画中游组群总平面图
Site Plan of the *Huazhongyou* Complex

0 1　　5　　　10m

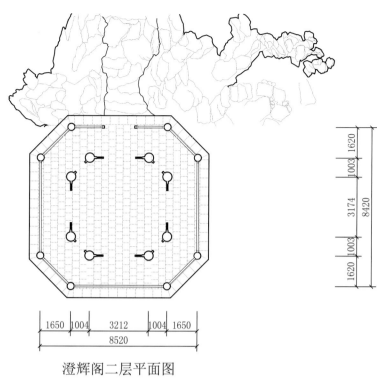

澄辉阁二层平面图

Second Floor Plan of the *Chenghui* Pavilion

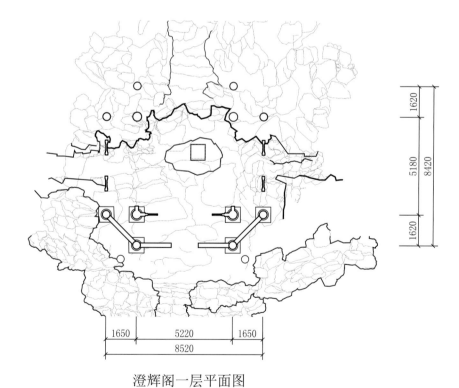

澄辉阁一层平面图

First Floor Plan of the *Chenghui* Pavilion

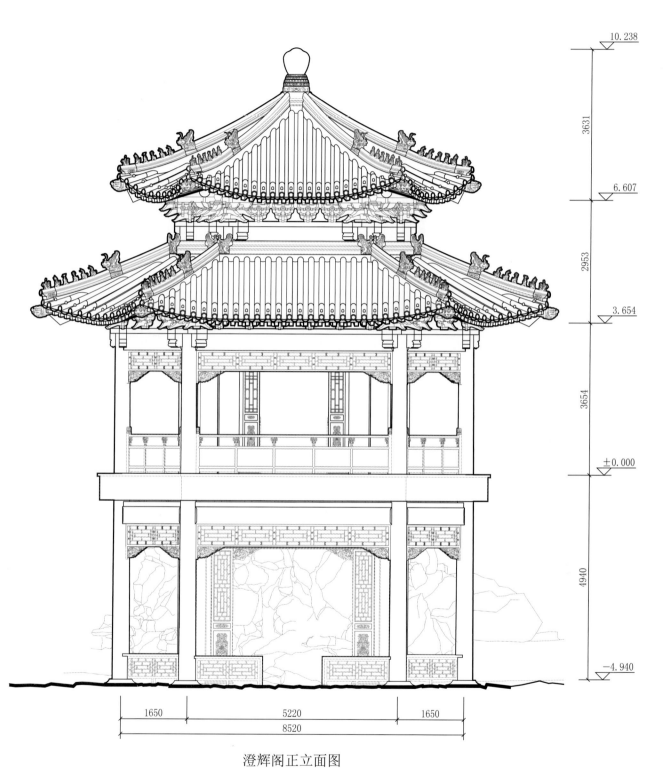

澄辉阁正立面图

Front Elevation of the *Chenghui* Pavilion

画中游正立面图
Front Elevation of the *Huazhongyou*

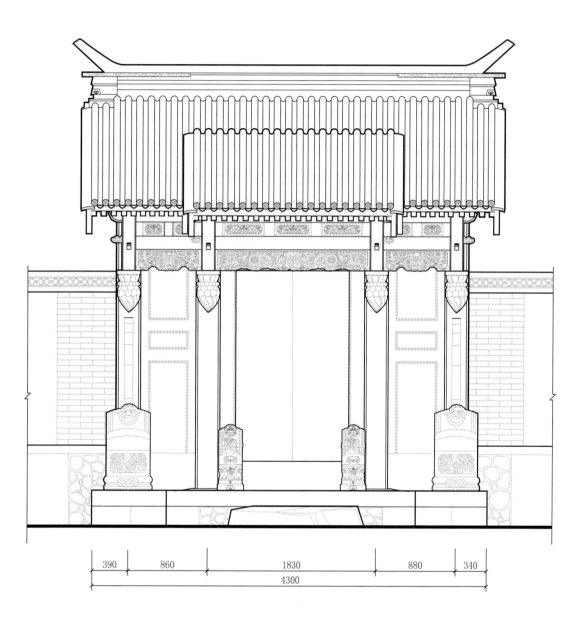

画中游垂花门正立面图
Front Elevation of the Festooned Door of *Huazhongyou*

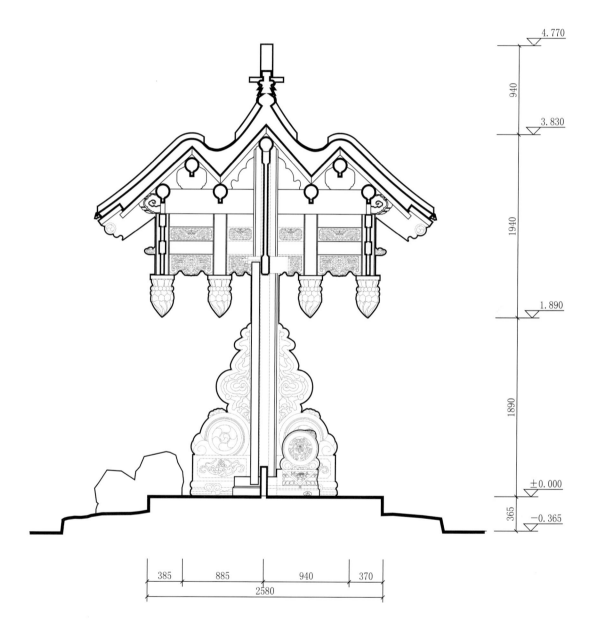

画中游垂花门剖面图
Section of the Festooned Door of *Huazhongyou*

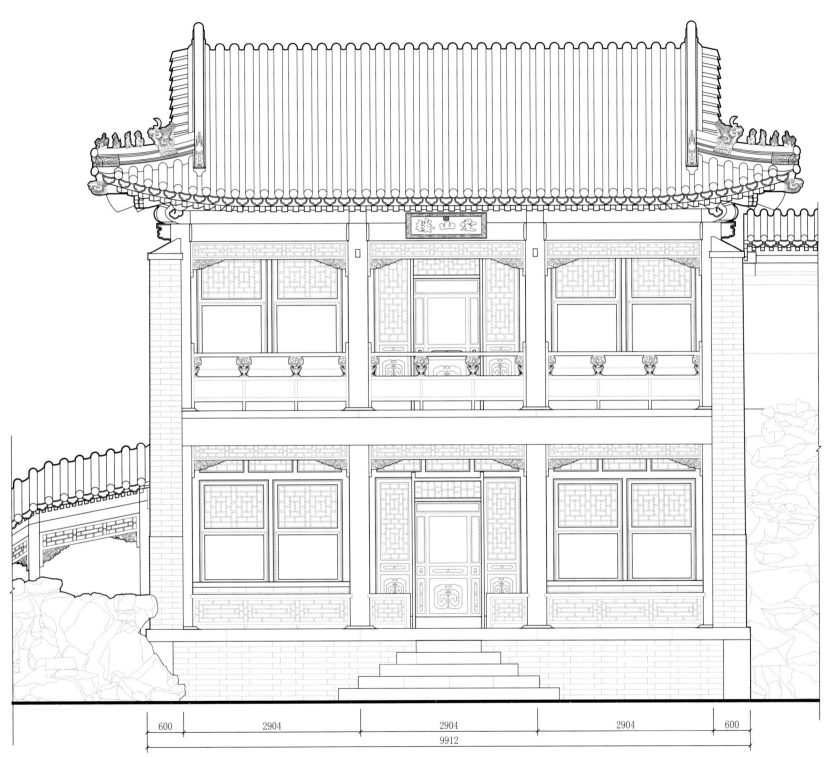

9.621
9.283
2574
6.709
834
5.875
2410
3.465
3465
±0.000
1140
−1.140

600 2904 2904 2904 600
9912

爱山楼正立面图
Front Elevation of the *Aishan* Pavilion

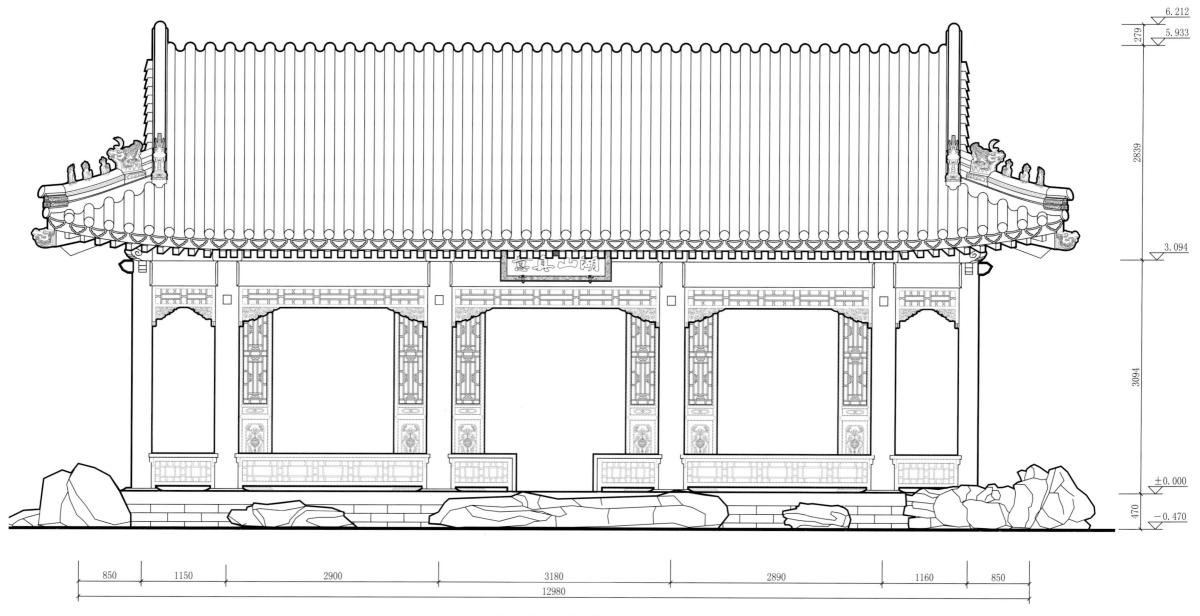

6.212
5.933
279
2839
3.094
3094
±0.000
470
-0.470

850 1150 2900 3180 2890 1160 850
12980

湖山真意正立面图
Front Elevation of the *Hushanzhenyi*

扬仁风组群总平面图（1956年测绘）

Site Plan of the *Yangrenfeng* Complex（Drawn in 1956）

扬仁风组群总平面图

Site Plan of the *Yangrenfeng* Complex

N

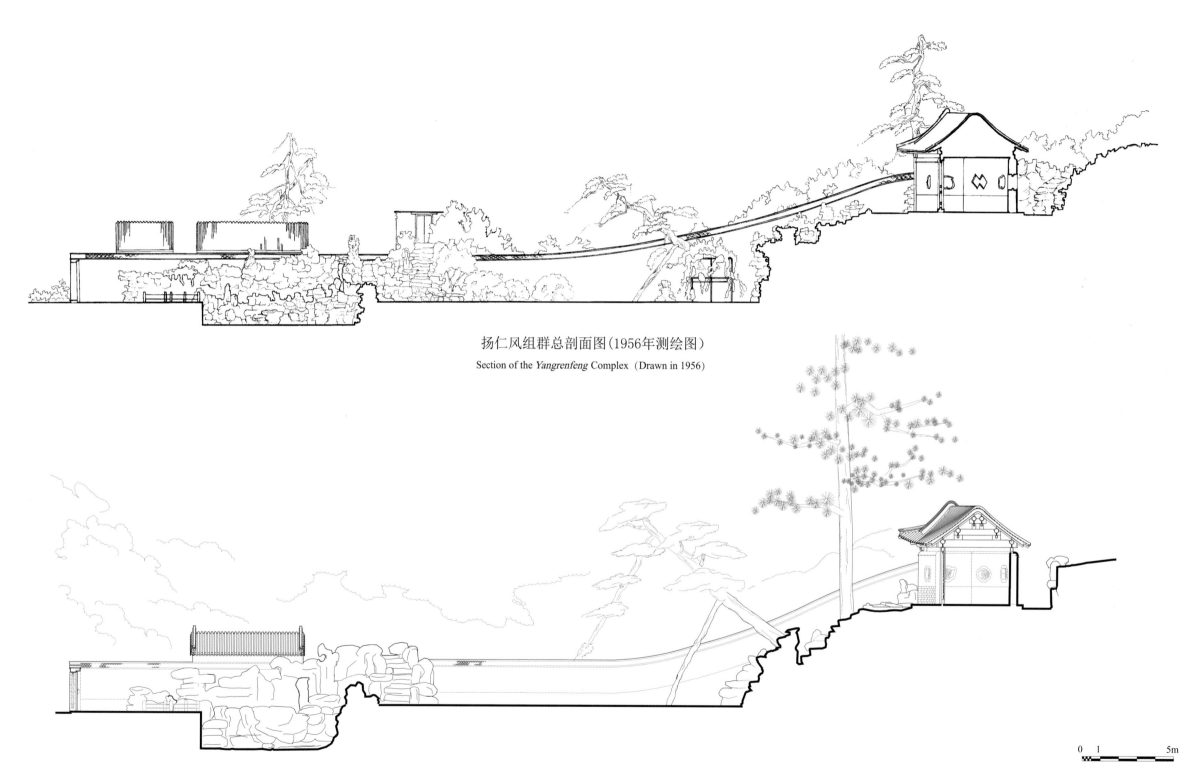

扬仁风组群总剖面图（1956年测绘图）

Section of the *Yangrenfeng* Complex （Drawn in 1956）

扬仁风组群总剖面图

Section of the *Yangrenfeng* Complex

0　1　　　　5m

扬仁风正立面图（1956 年测绘图）

Front Elevation of the *Yangrenfeng* (Drawn in 1956)

扬仁风背立面图（1956 年测绘图）

Back Elevation of the *Yangrenfeng* (Drawn in 1956)

0   0.5   1m

1 南房 South Hall
2 西房 West Hall
3 北房 North Hall
4 六角亭 Hexagonal Pavilion
5 八角亭 Octagonal Pavilion
6 东配殿 East Side Hall
7 西配殿 West Side Hall
8 畅观堂 *Changguan* Hall
9 睇佳榭 *Dijia* Hall

畅观堂组群总平面图
Site Plan of the *Changguan* Hall Complex

0　1　　　5m

畅观堂组群1-1剖面图

1-1 Section of the *Changguan* Hall Complex

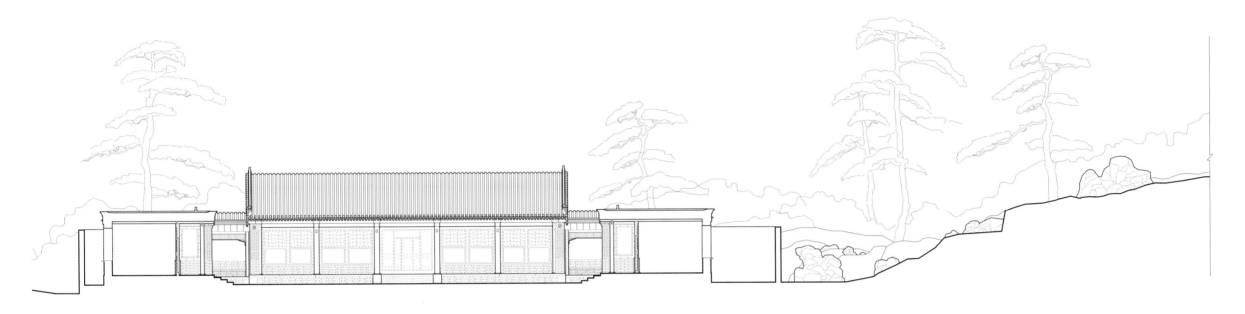

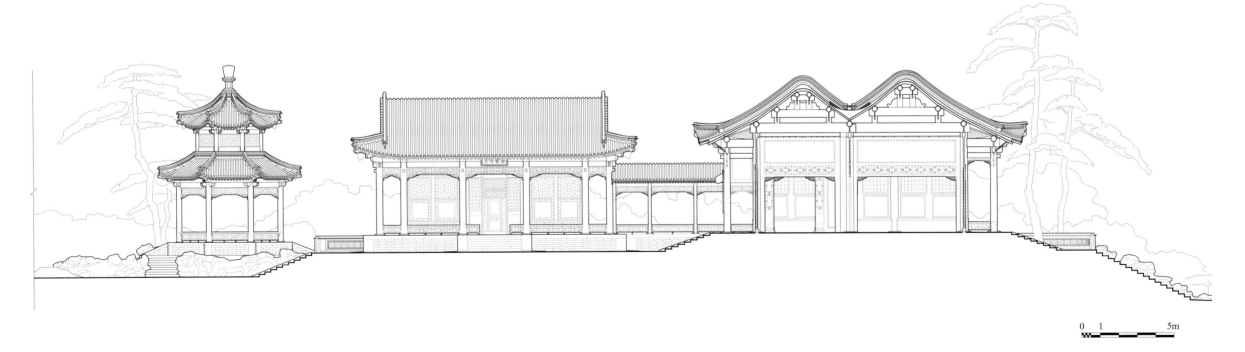

畅观堂组群2-2剖面图

2-2 Section of the *Changguan* Hall Complex

| 1620 | 3330 | 3480 | 3840 | 3480 | 3330 | 1620 |

20700

畅观堂正立面图

Front Elevation of the *Changguan* Hall

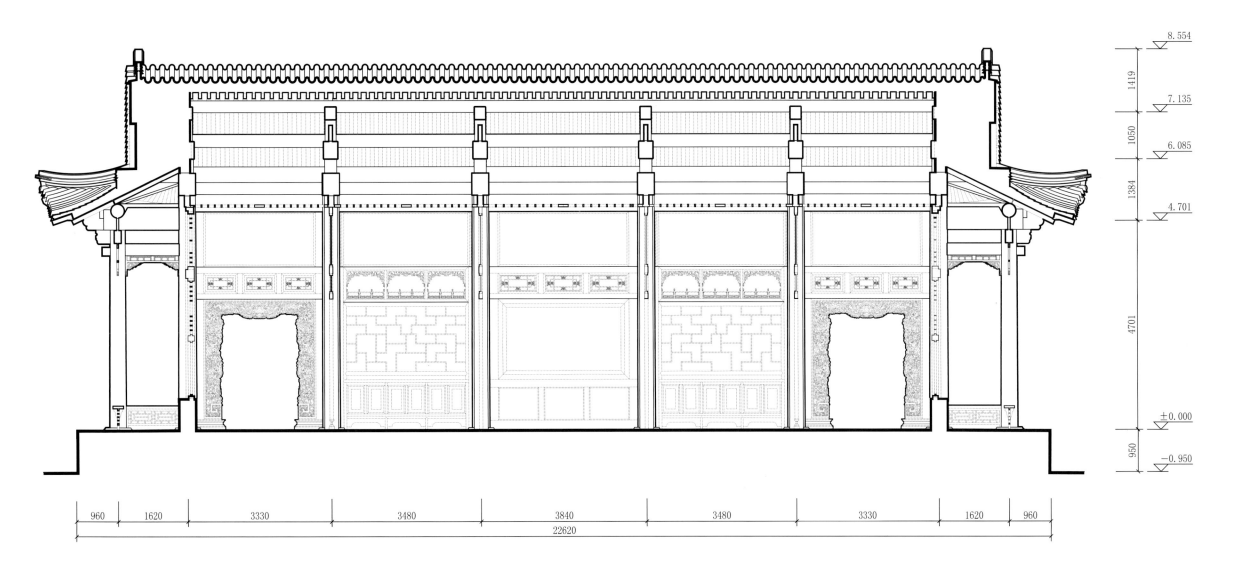

8.554

1419

7.135

1050

6.085

1384

4.701

4701

±0.000

950

−0.950

| 960 | 1620 | 3330 | 3480 | 3840 | 3480 | 3330 | 1620 | 960 |

22620

畅观堂纵剖面图
Longitudinal Section of the *Changguan* Hall

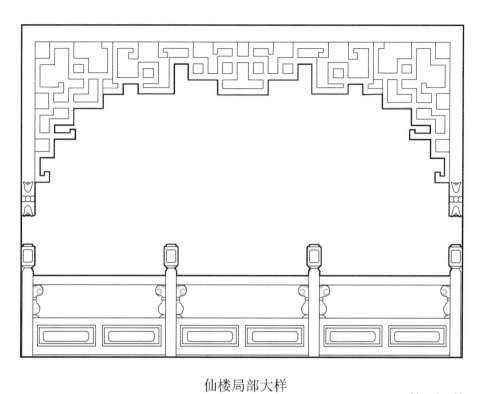

仙楼局部大样
Detail Drawing of *Xianlou*

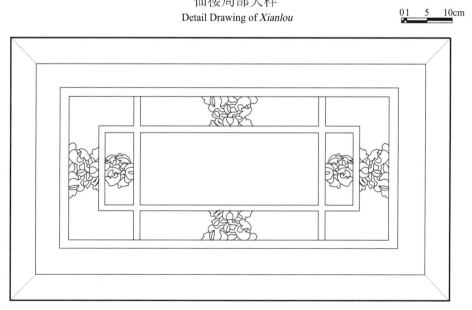

畅观堂门窗大样
Window of *Changguan* Hall

东梢间落地罩立面图
Elevation of the *Luodizhao* of East Shaojian

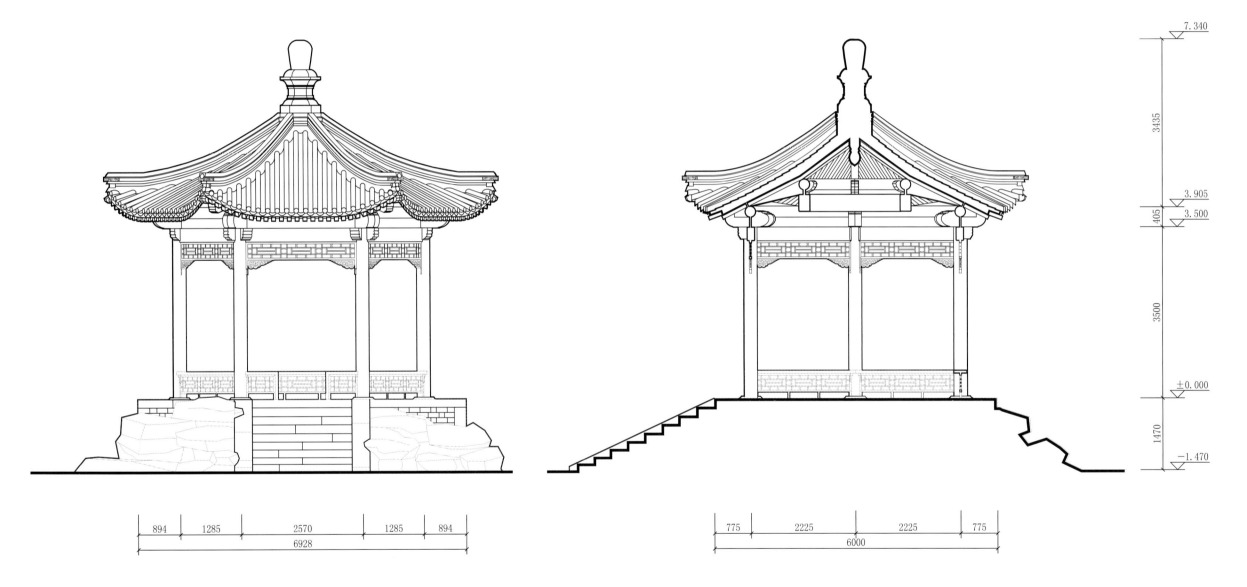

六角亭正立面图
Front Elevation of the Hexagonal Pavilion

六角亭剖面图
Section of the Hexagonal Pavilion

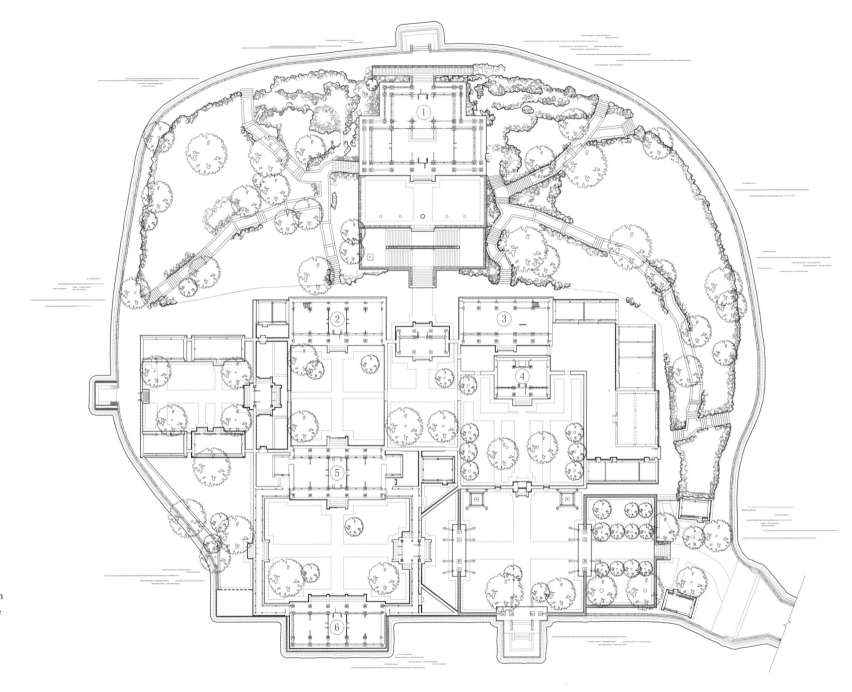

1 涵虚堂 *Hanxu* Hall
2 月波楼 *Yuebo* Paviloin
3 云香阁 *Yunxiang* Pavilion
4 广润祠 *Guangrun* Temple
5 澹会轩 *Danhui* Hall
6 鉴远堂 *Jianyuan* Hall

南湖岛组群总平面图
Site Plan of the South Lake Island Complex

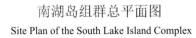

N

01 5 10m

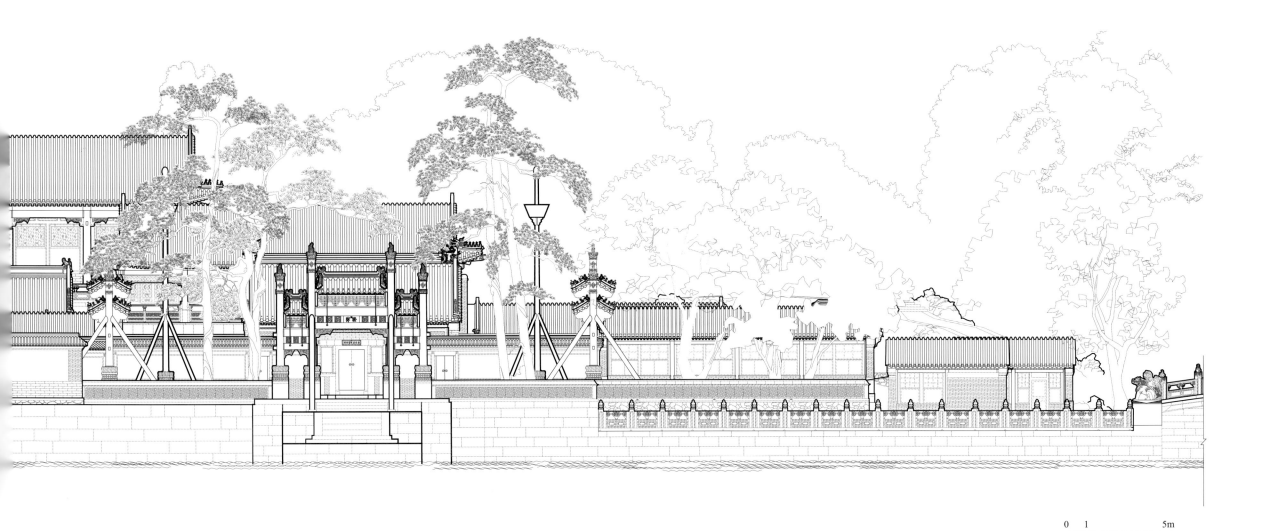

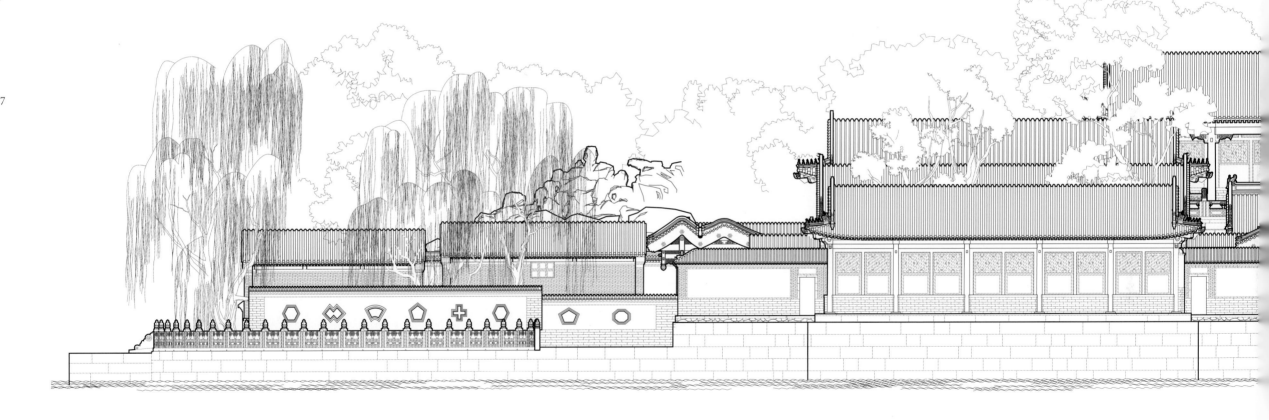

南湖岛组群正立面图

Front Elevation of the South Lake Island Complex

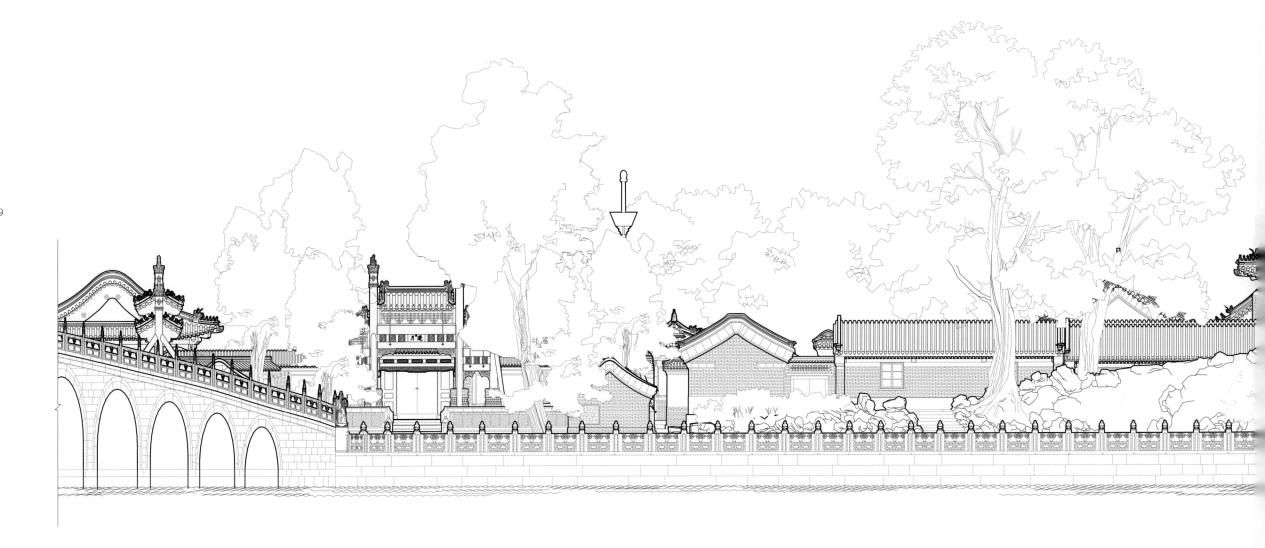

南湖岛组群侧立面图

Side Elevation of the South Lake Island Complex

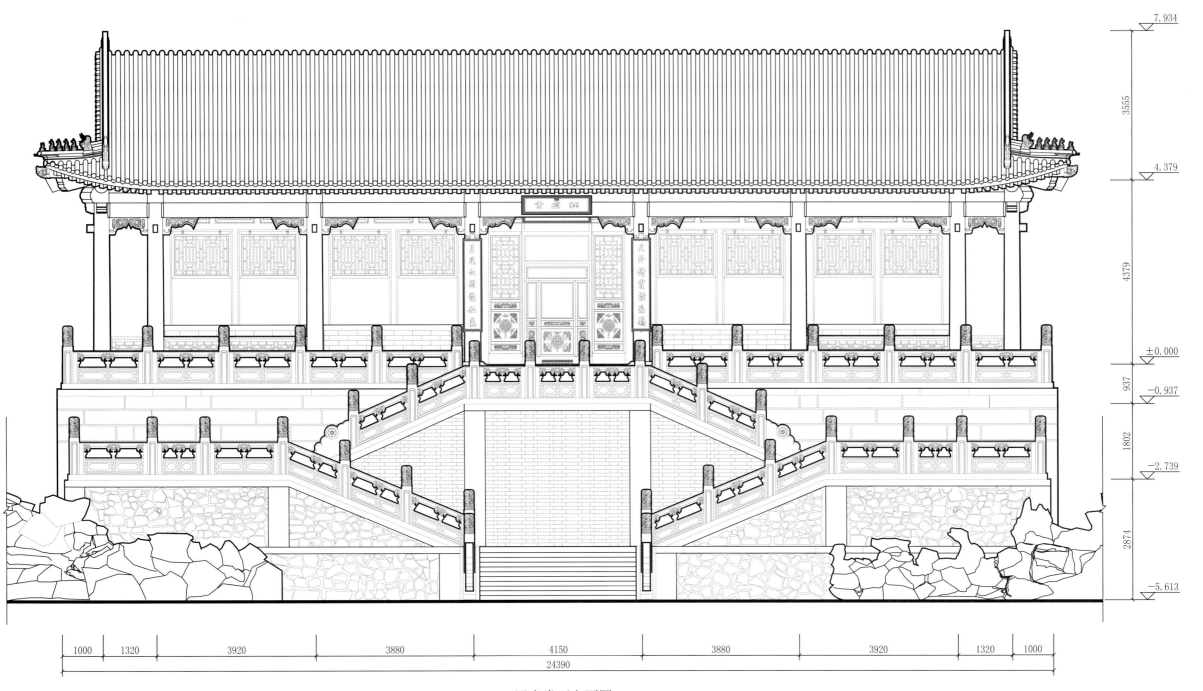

7.934

3555

4.379

4379

±0.000

937 −0.937

1802

−2.739

2874

−5.613

1000 1320 3920 3880 4150 3880 3920 1320 1000

24390

涵虚堂正立面图
Front Elevation of the *Hanxu* Hall

涵虚堂明间剖面图

Mingjian Section of the *Hanxu* Hall

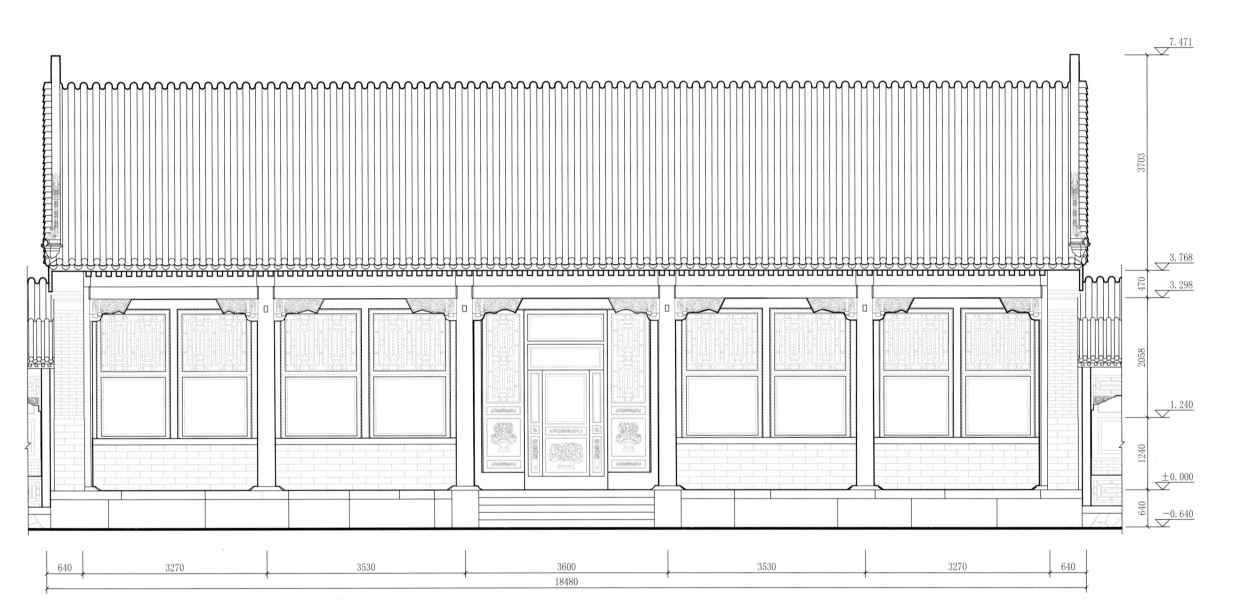

7.471

3703

3.768

470

3.298

232

2058

1.240

1240

±0.000

640

−0.640

640　3270　3530　3600　3530　3270　640

18480

澹会轩正立面图

Front Elevation of the *Danhui* Pavilion

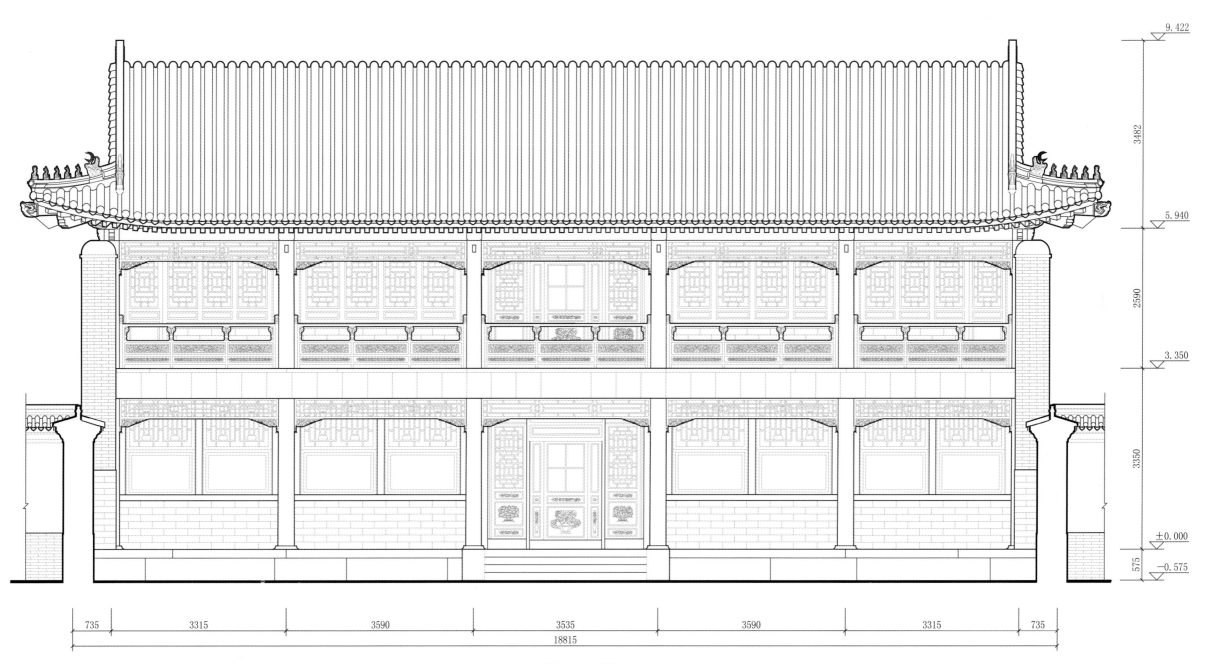

9.422

3482

5.940

2590

3.350

3350

±0.000

575

−0.575

735　3315　3590　3535　3590　3315　735

18815

月波楼正立面图

Front Elevation of the *Yuebo* Pavilion

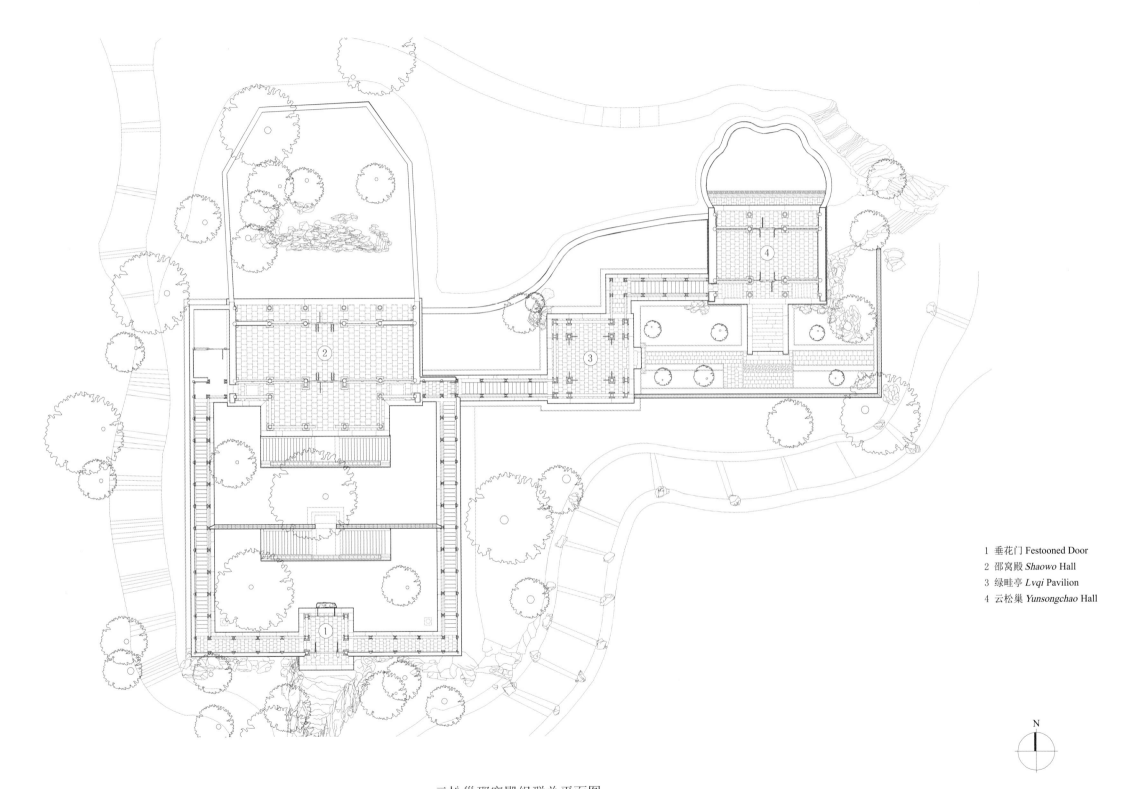

1　垂花门 Festooned Door
2　邵窝殿 *Shaowo* Hall
3　绿畦亭 *Lvqi* Pavilion
4　云松巢 *Yunsongchao* Hall

N

云松巢邵窝殿组群总平面图
Site Plan of the *Yunsongchao* Hall and *Shaowo* Hall Complex

0　1　　5　　10m

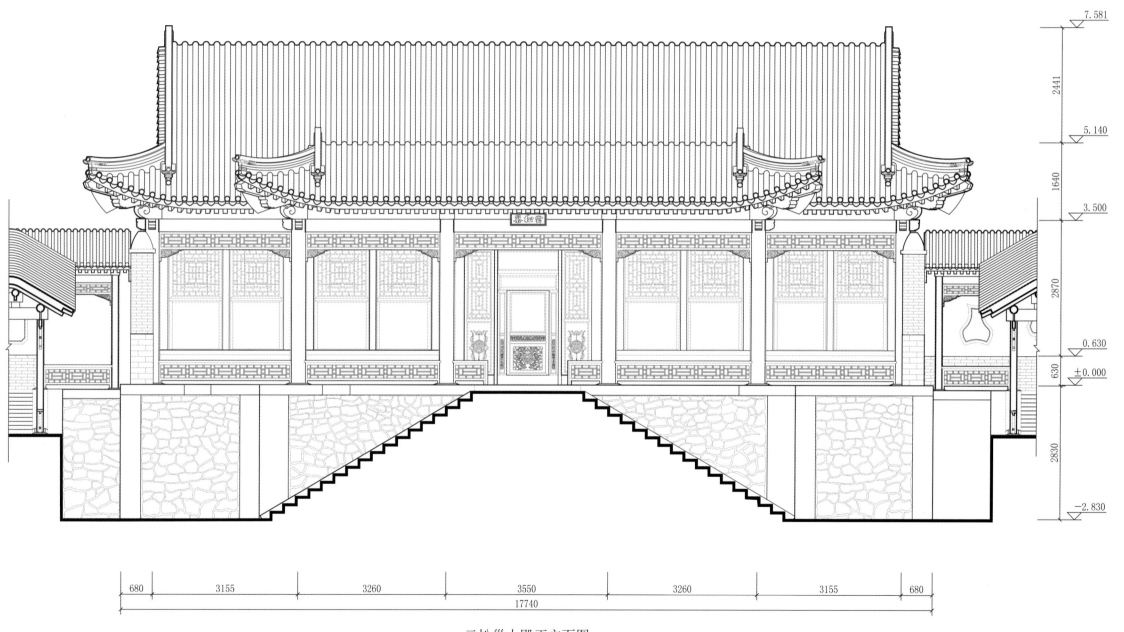

7.581

2441

5.140

1640

3.500

2870

0.630

630   ±0.000

2830

-2.830

680   3155   3260   3550   3260   3155   680

17740

云松巢大殿正立面图

Front Elevation of the *Yunsongchao* Hall

0　1　　　　　5m

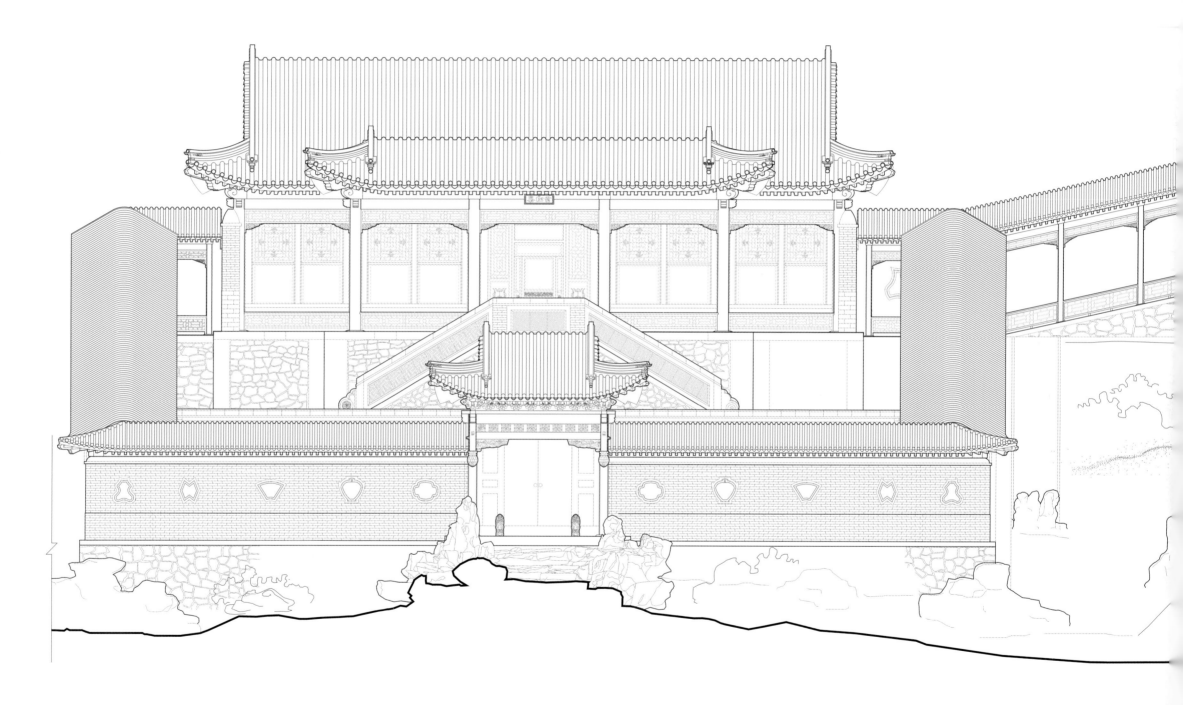

云松巢邵窝殿组群正立面图

Front Elevation of the *Yunsongchao* Hall and *Shaowo* Hall Complex

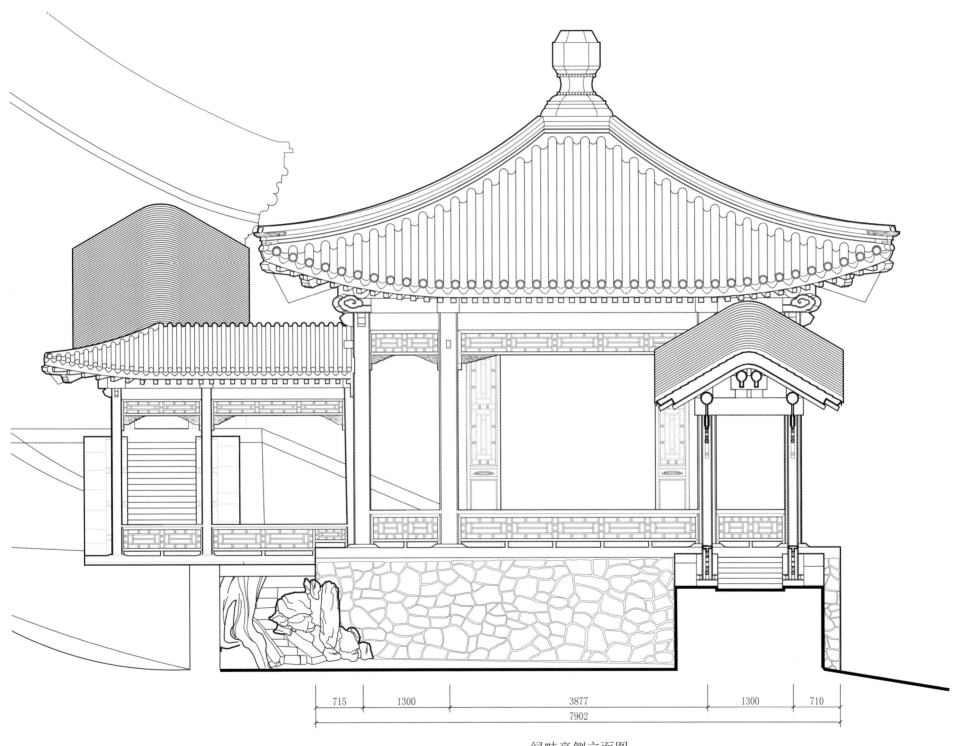

7.510

1243

6.267

2643

3.624

3624

±0.000

1762

−1.762

715　1300　3877　1300　710

7902

绿畦亭侧立面图

Side Elevation of the *Lvqi* Pavilion

邵窝殿正立面图

Front Elevation of the *Shaowo* Hall

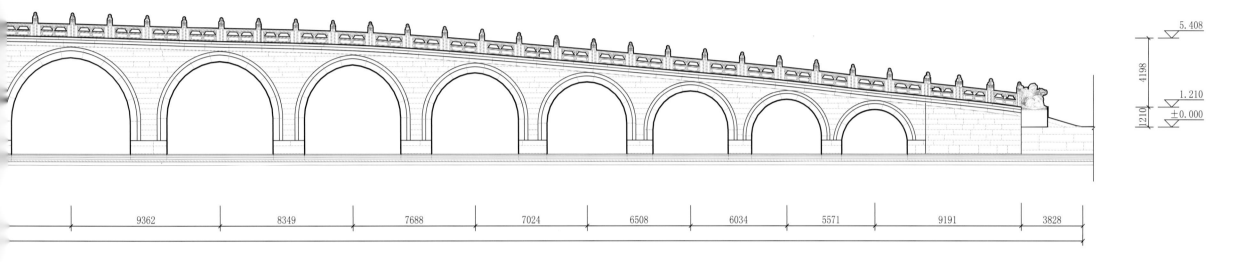

5.408

4198

1.210

±0.000

1210

9362　　8349　　7688　　7024　　6508　　6034　　5571　　9191　　3828

240

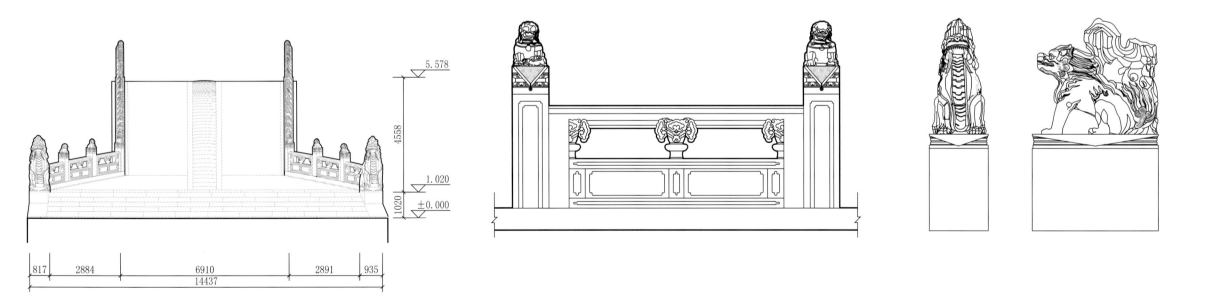

5.578

4558

1.020

±0.000

1020

817　2884　　6910　　2891　935
14437

十七孔桥侧立面图
Side Elevation of the Seventeen-arch Bridge

十七孔桥栏板大样图
Handrail of the Seventeen-arch Bridge

十七孔桥靠山兽大样图
Kaoshanshou of the Seventeen-arch Bridge

0　　　0.5　　　1m

0　　　0.5　　　1m

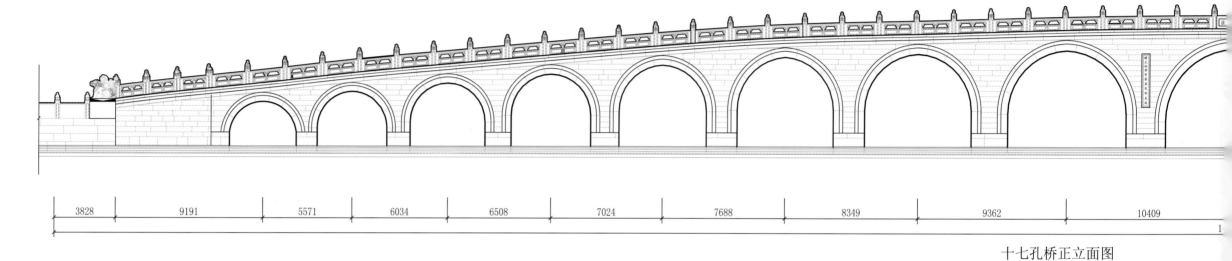

| 3828 | 9191 | 5571 | 6034 | 6508 | 7024 | 7688 | 8349 | 9362 | 10409 |

十七孔桥正立面图

Front Elevation of the Seventeen-arch Bridge

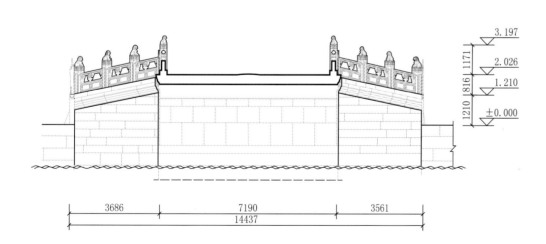

十七孔桥剖面图 1

1 Section of the Seventeen-arch Bridge

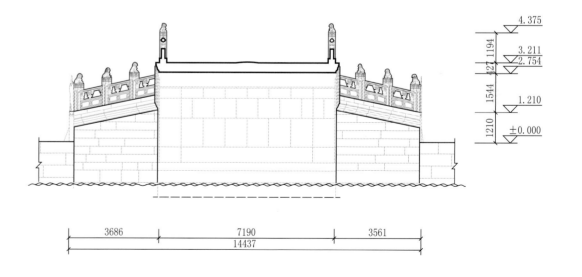

十七孔桥剖面图 2

2 Section of the Seventeen-arch Bridge

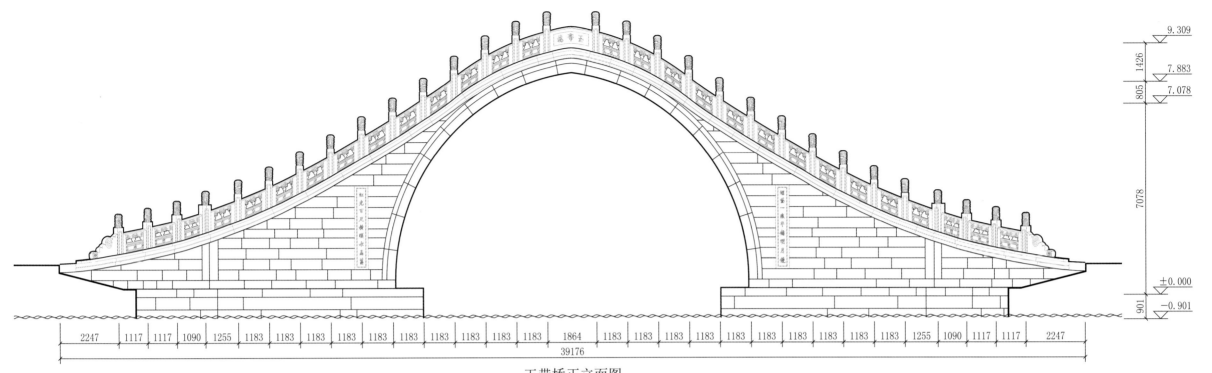

玉带桥正立面图

Front Elevation of the *Yudai* Bridge

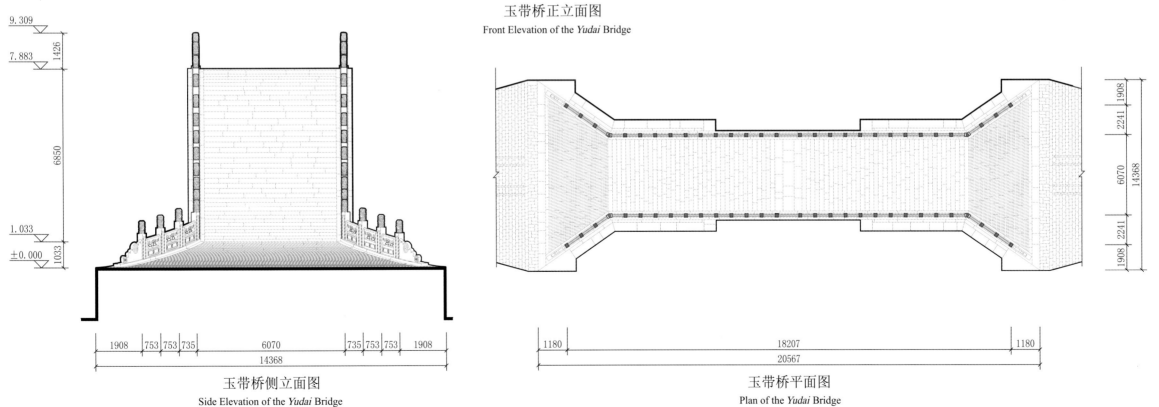

玉带桥侧立面图

Side Elevation of the *Yudai* Bridge

玉带桥平面图

Plan of the *Yudai* Bridge

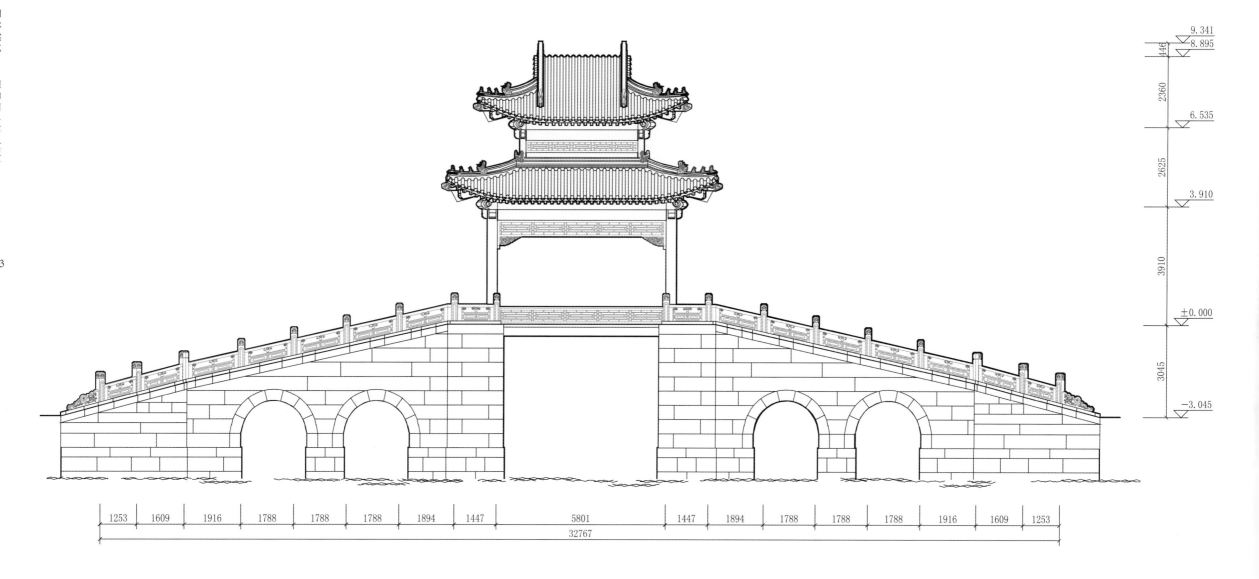

柳桥正立面图

Front Elevation of the *Liu* Bridge

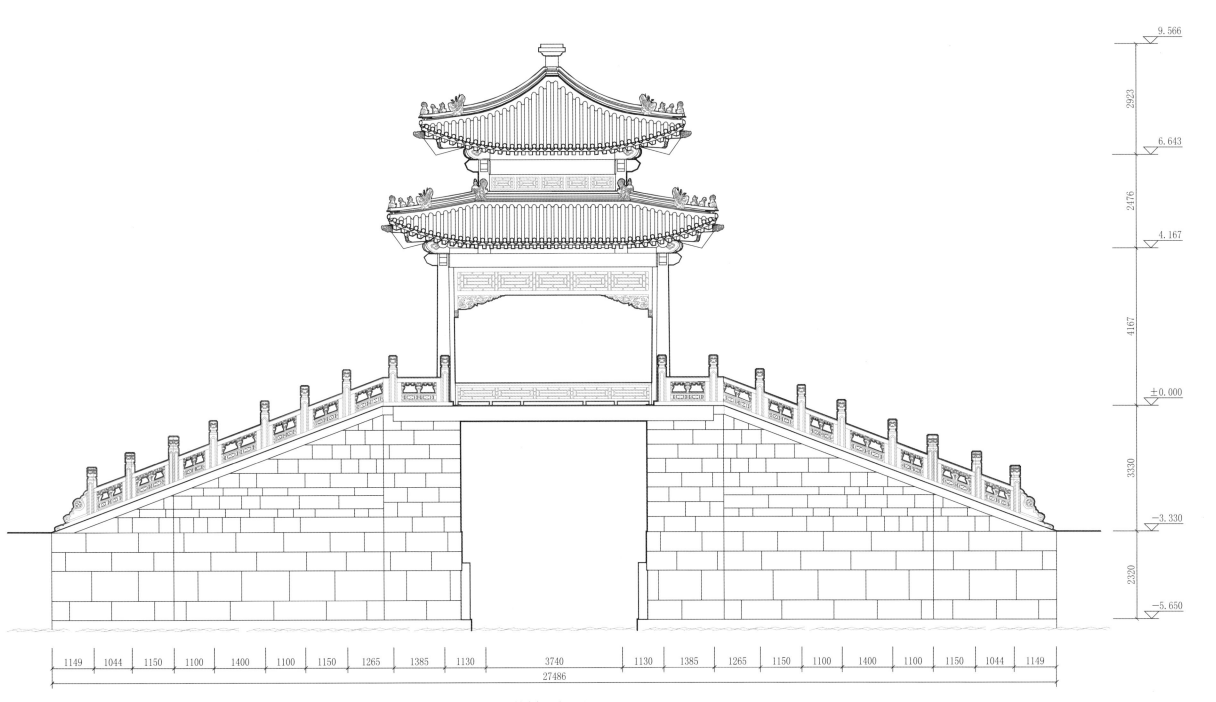

9.566

2923

6.643

2476

4.167

4167

±0.000

3330

−3.330

2320

−5.650

| 1149 | 1044 | 1150 | 1100 | 1400 | 1100 | 1150 | 1265 | 1385 | 1130 | 3740 | 1130 | 1385 | 1265 | 1150 | 1100 | 1400 | 1100 | 1150 | 1044 | 1149 |

27486

练桥正立面图

Front Elevation of the *Lian* Bridge

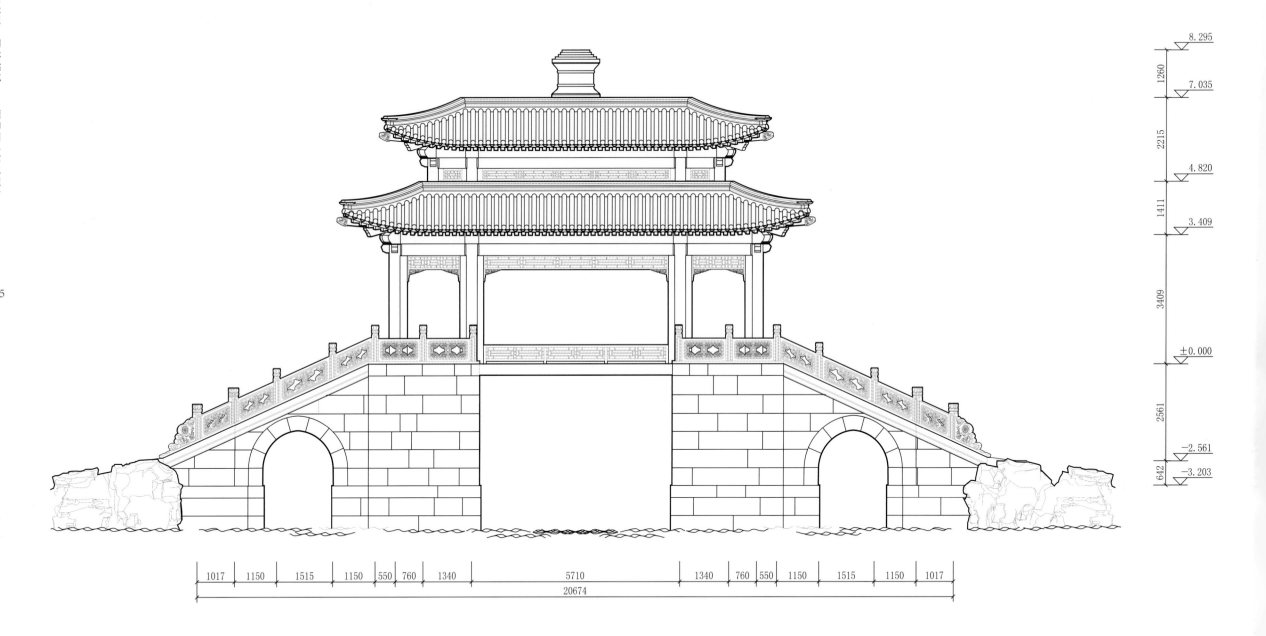

8.295

1260

7.035

2215

4.820

1411

3.409

3409

±0.000

2561

−2.561

642

−3.203

| 1017 | 1150 | 1515 | 1150 | 550 | 760 | 1340 | 5710 | 1340 | 760 | 550 | 1150 | 1515 | 1150 | 1017 |

20674

豳风桥正立面图

Front Elevation of the *Binfeng* Bridge

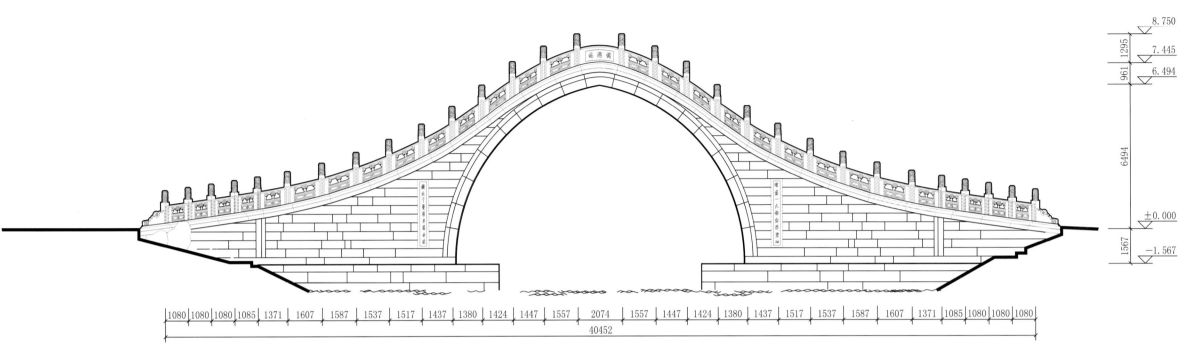

绣漪桥正立面图

Front Elevation of the *Xiuyi* Bridge

中国古建筑测绘大系·园林建筑——颐和园（第二版）

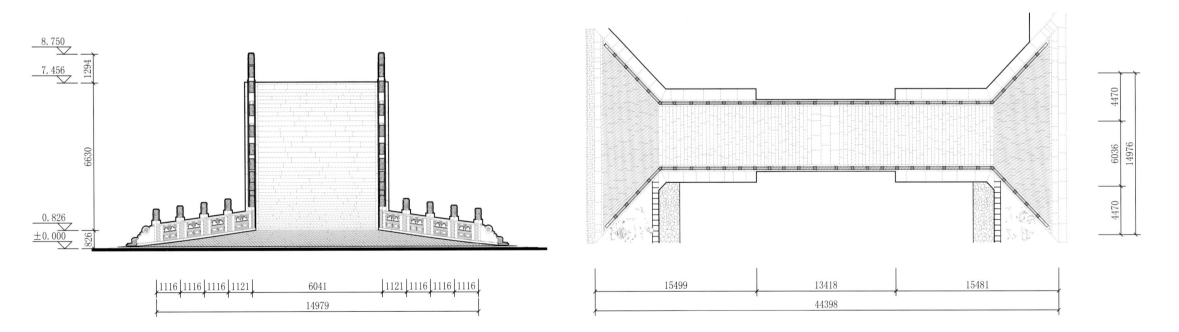

绣漪桥侧立面图

Side Elevation of the *Xiuyi* Bridge

绣漪桥平面图

Plan of the *Xiuyi* Bridge

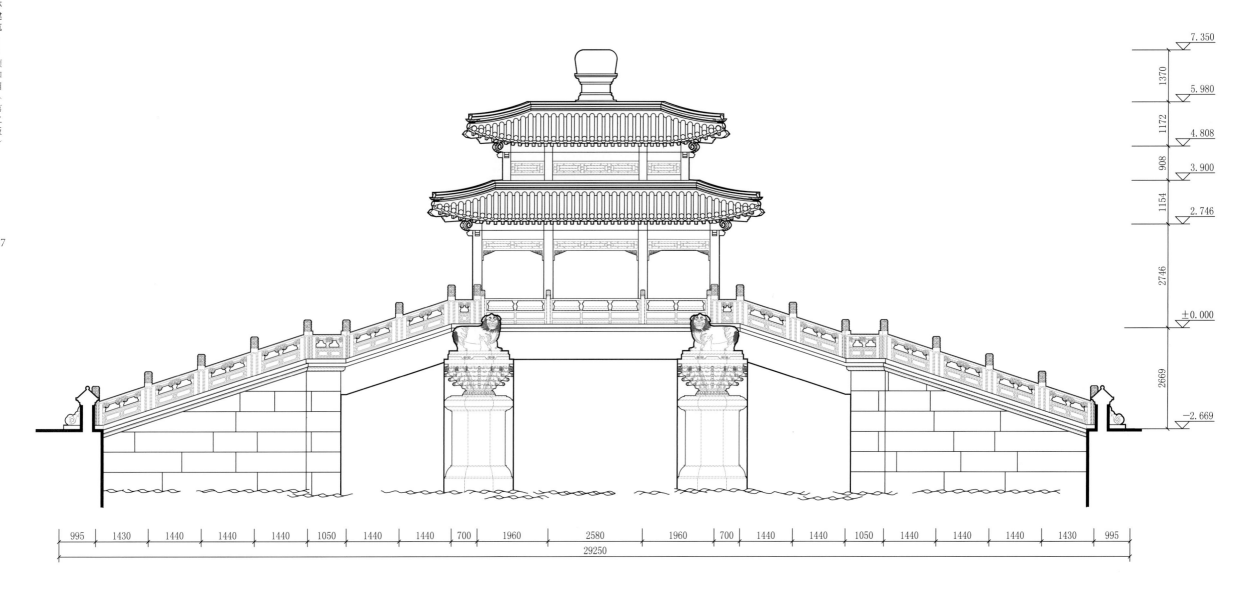

荇桥正立面图
Front Elevation of the *Xing* Bridge

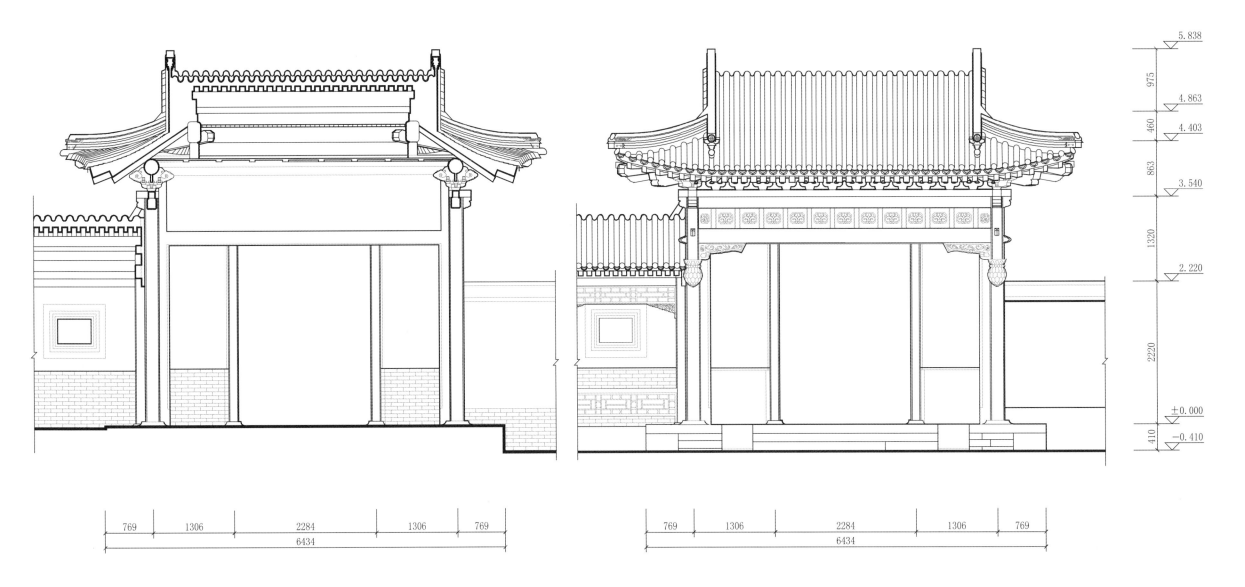

邀月门剖面图
Section of the *Yaoyue* Gate

邀月门正立面图
Front Elevation of the *Yaoyue* Gate

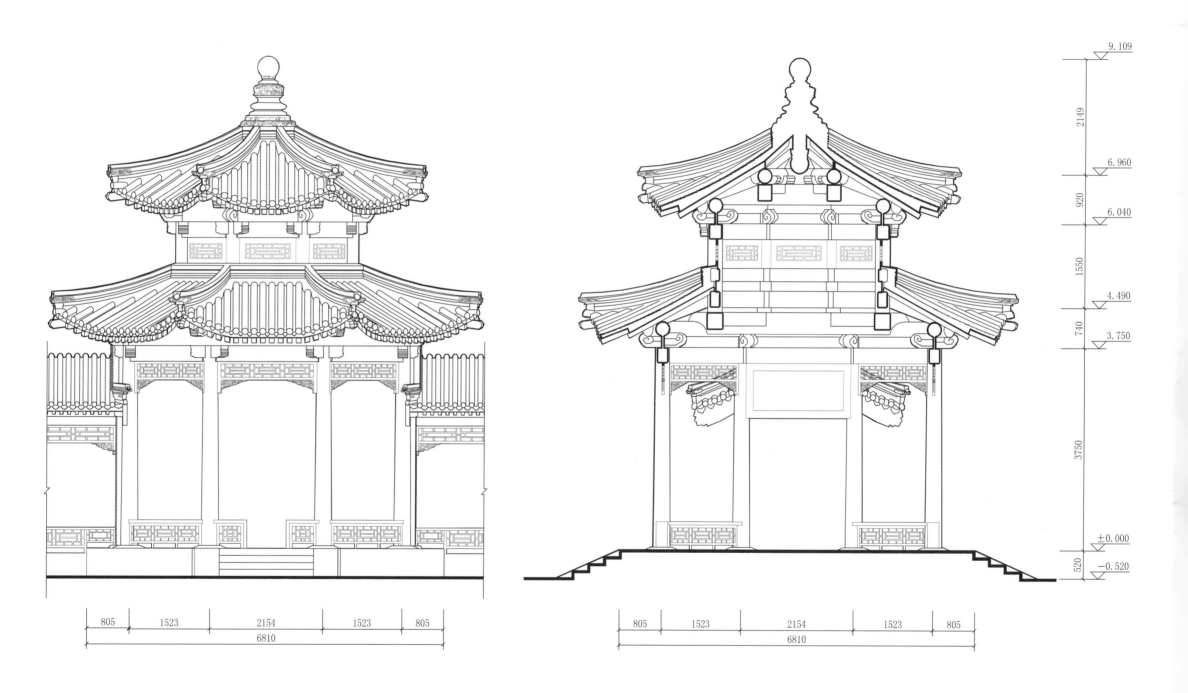

9.109

2149

6.960

920

6.040

1550

4.490

740

3.750

3750

±0.000

520

−0.520

805 | 1523 | 2154 | 1523 | 805
6810

805 | 1523 | 2154 | 1523 | 805
6810

清遥亭正立面图
Front Elevation of the *Qingyao* Pavilion

清遥亭剖面图
Section of the *Qingyao* Pavilion

807 954 660 393 1497 393 660 954 807
7125

秋水亭正立面图
Front Elevation of the *Qiushui* Pavilion

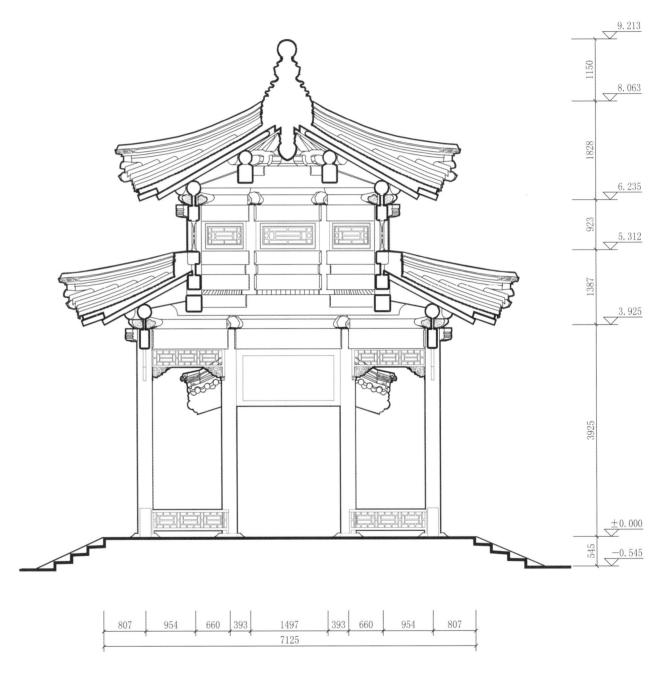

9.213
1150
8.063
1828
6.235
923
5.312
1387
3.925
3925
±0.000
545
−0.545

250

807 954 660 393 1497 393 660 954 807
7125

秋水亭剖面图
Section of the *Qiushui* Pavilion

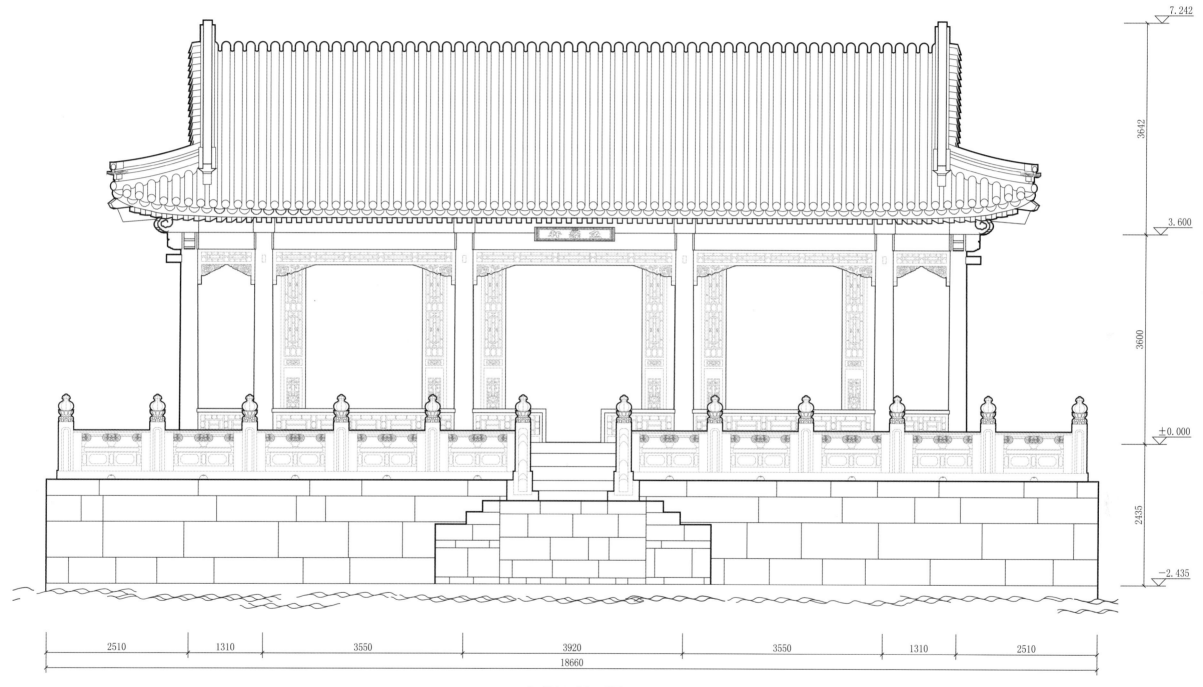

7.242

3642

3.600

3600

±0.000

2435

−2.435

2510 1310 3550 3920 3550 1310 2510

18660

鱼藻轩正立面图

Front Elevation of the *Yuzao* Pavilion

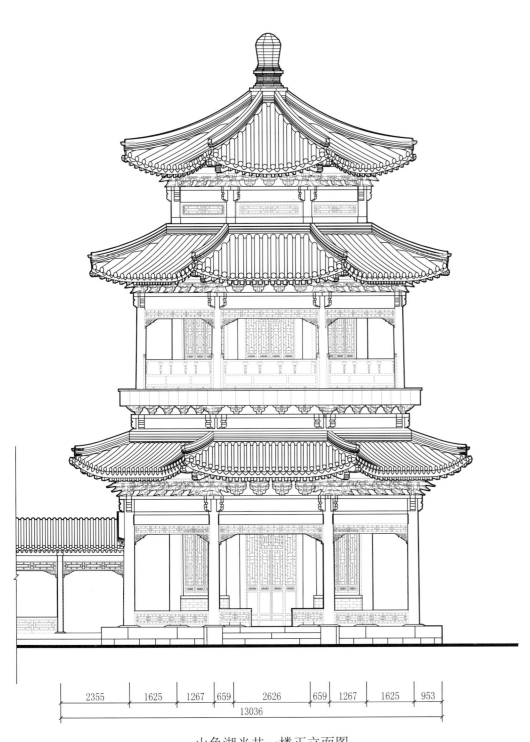

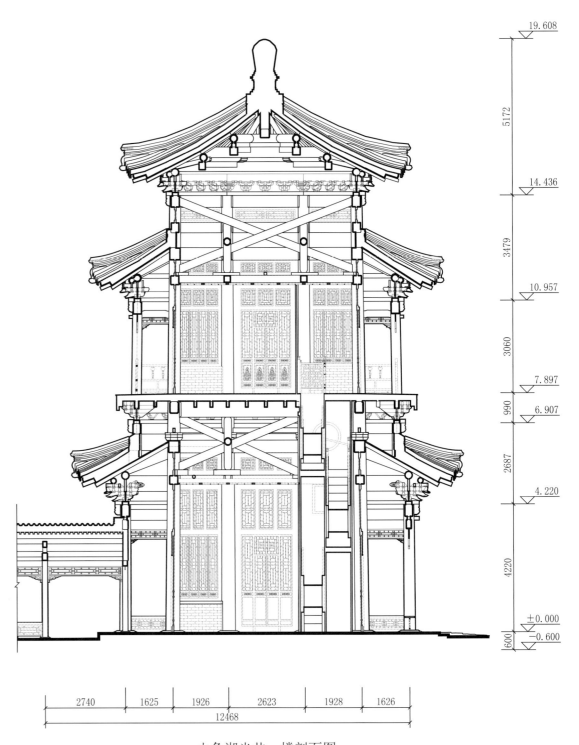

山色湖光共一楼正立面图
Front Elevation of the *Shansehuguang-gongyilou* Pavilion

山色湖光共一楼剖面图
Section of the *Shansehuguang-gongyilou* Pavilion

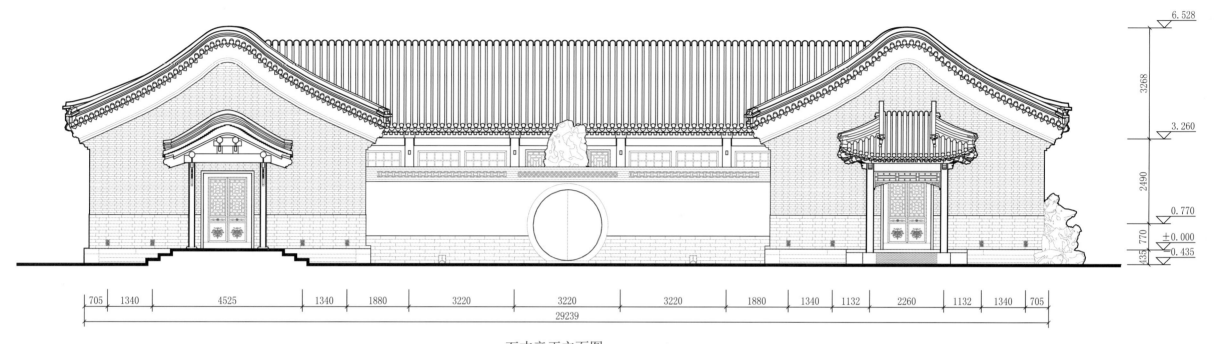

石丈亭正立面图

Front Elevation of the *Shizhang* Pavilion

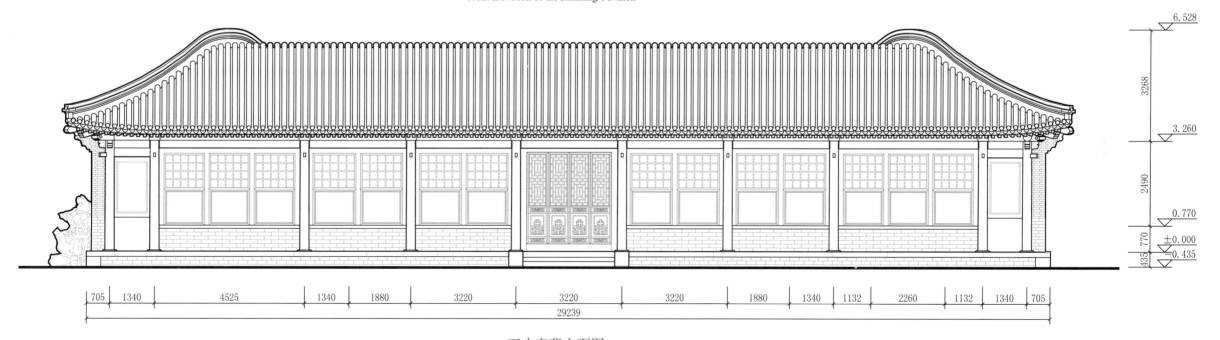

石丈亭背立面图

Back Elevation of the *Shizhang* Pavilion

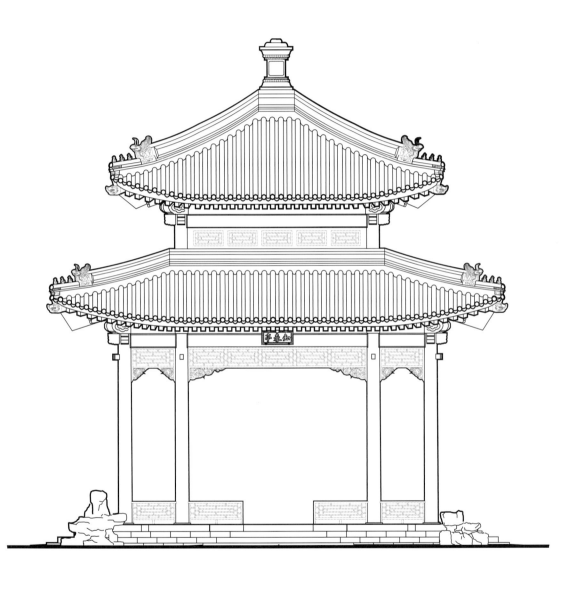

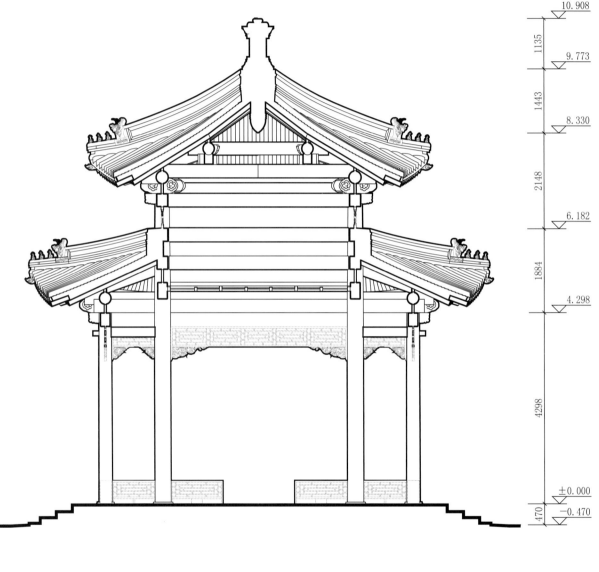

| 805 | 1330 | 4495 | 1330 | 805 |

8765

| 805 | 1330 | 4495 | 1330 | 805 |

8765

10.908
1135
9.773
1443
8.330
2148
6.182
1884
4.298
4298
±0.000
470
-0.470

知春亭正立面图

Front Elevation of the *Zhichun* Pavilion

知春亭剖面图

Section of the *Zhichun* Pavilion

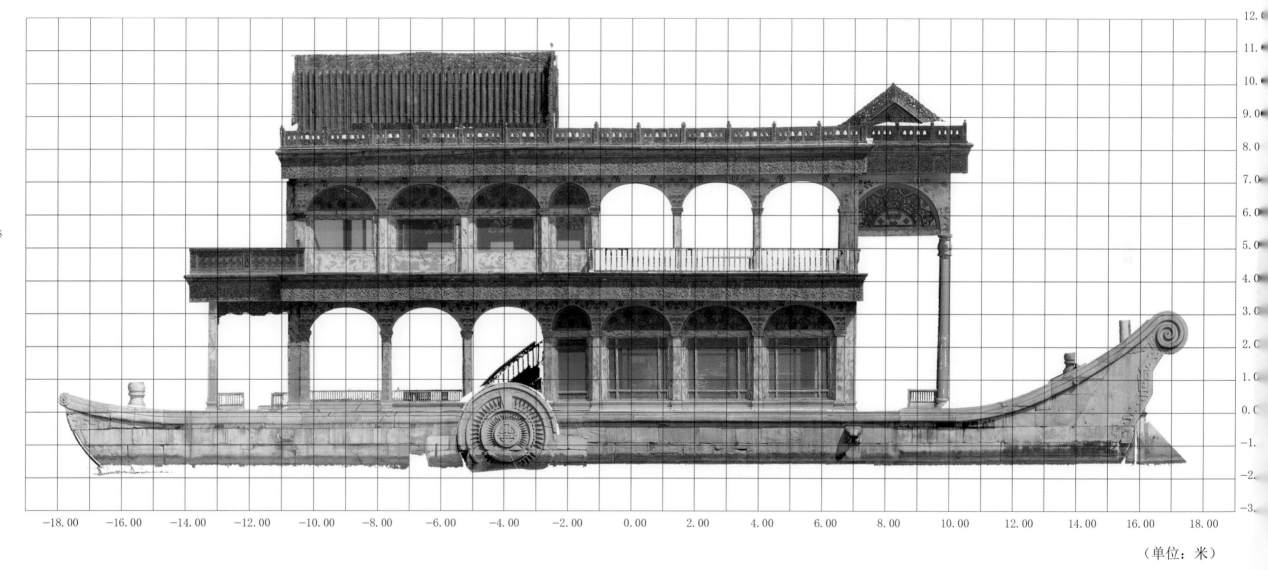

石舫正立面摄影像

Orthophoto of the Front Elevation of the Marble Boat

（单位：米）

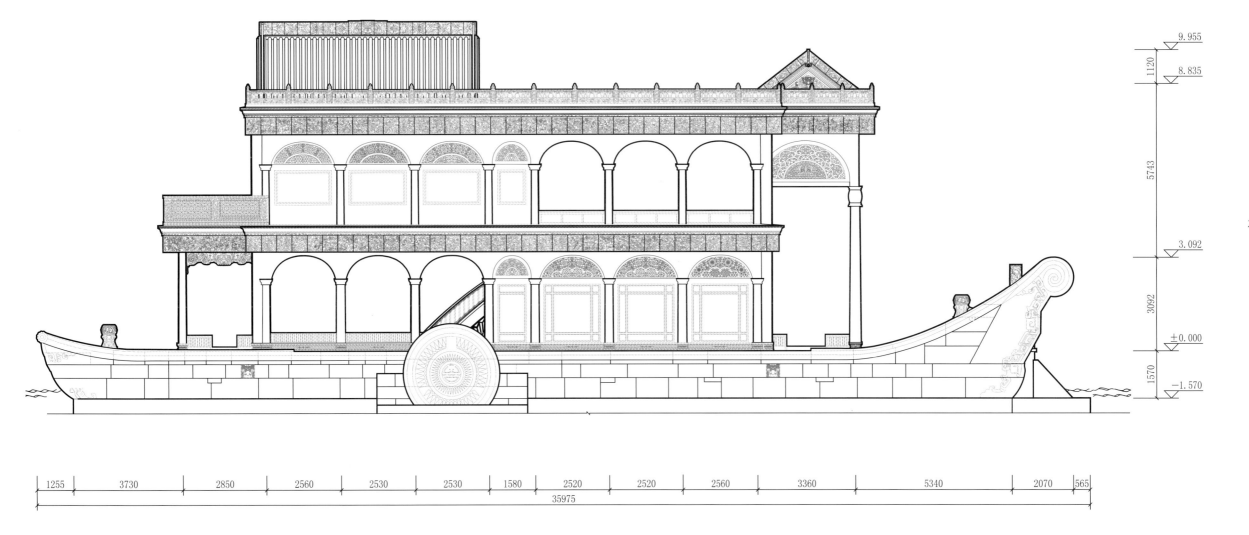

石舫正立面图
Front Elevation of the Marble Boat

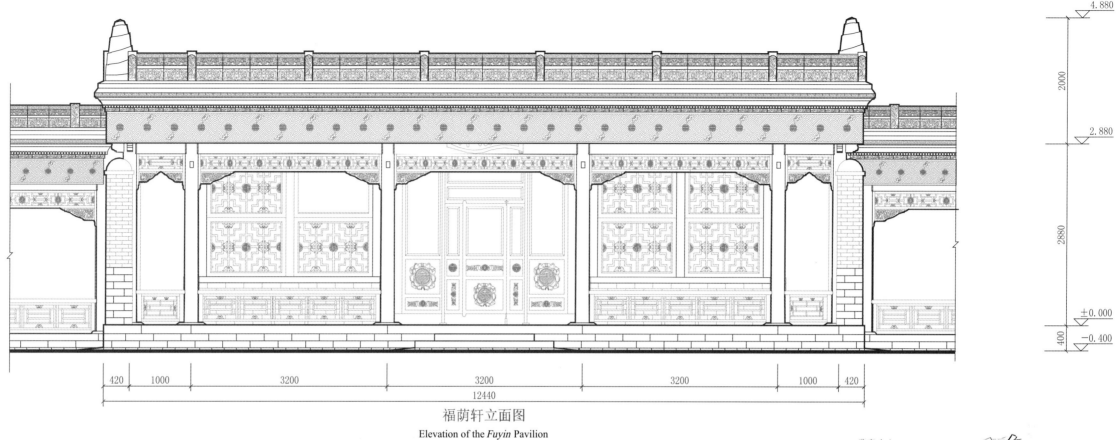

福荫轩立面图

Elevation of the *Fuyin* Pavilion

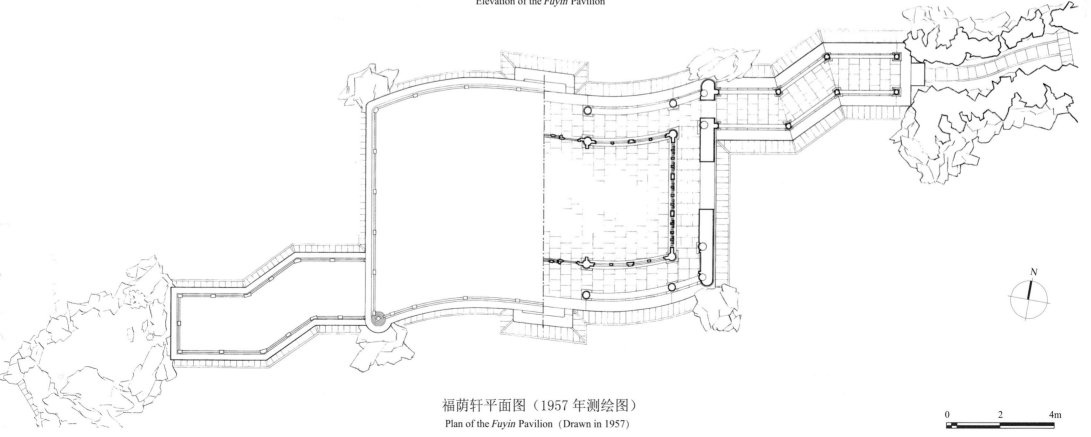

福荫轩平面图（1957 年测绘图）

Plan of the *Fuyin* Pavilion （Drawn in 1957）

0　　　2　　　4m

2309　418　491　491　491　418

482　284　283　283　283　283　283
2464

荟亭梁架仰视图
Framework Bottom View of the *Hui* Pavilion

2309　418　491　491　491　418

283　283　283　283　283　283　283
1981

荟亭梁架俯视图
Framework Top View of the *Hui* Pavilion

4.930
615
4.315
1755
2.560
2050
0.510
510　±0.000
583　−0.583

732　699　850　1700　850　1700　850　699　732
8812

荟亭正立面图
Front Elevation of the *Hui* Pavilion

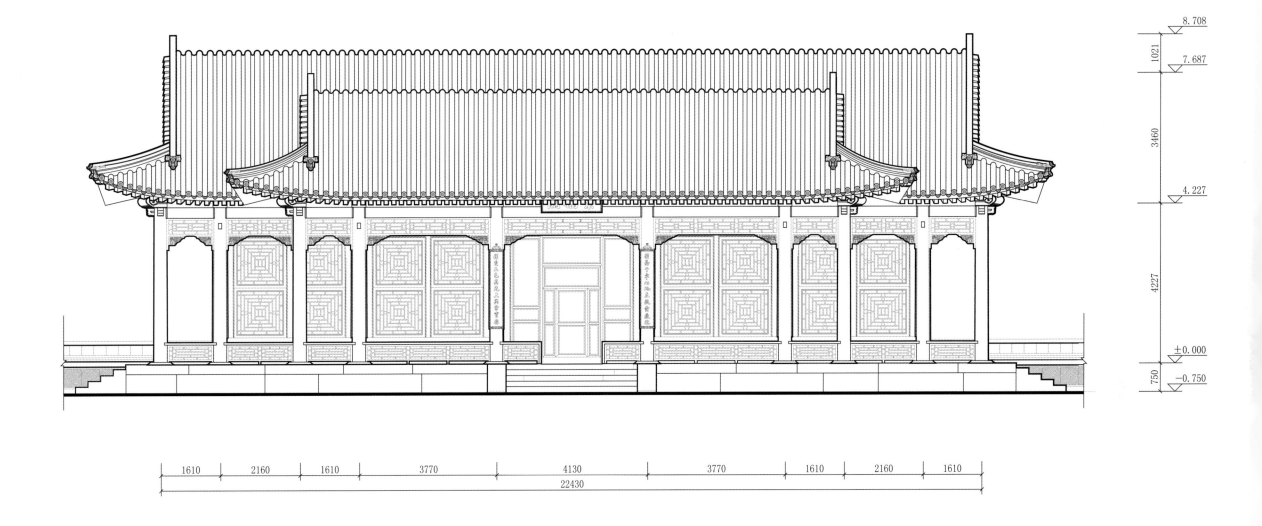

8.708

1021

7.687

3460

4.227

4227

±0.000

750

−0.750

1610 2160 1610 3770 4130 3770 1610 2160 1610

22430

景福阁正立面图

Front Elevation of the *Jingfu* Pavilion

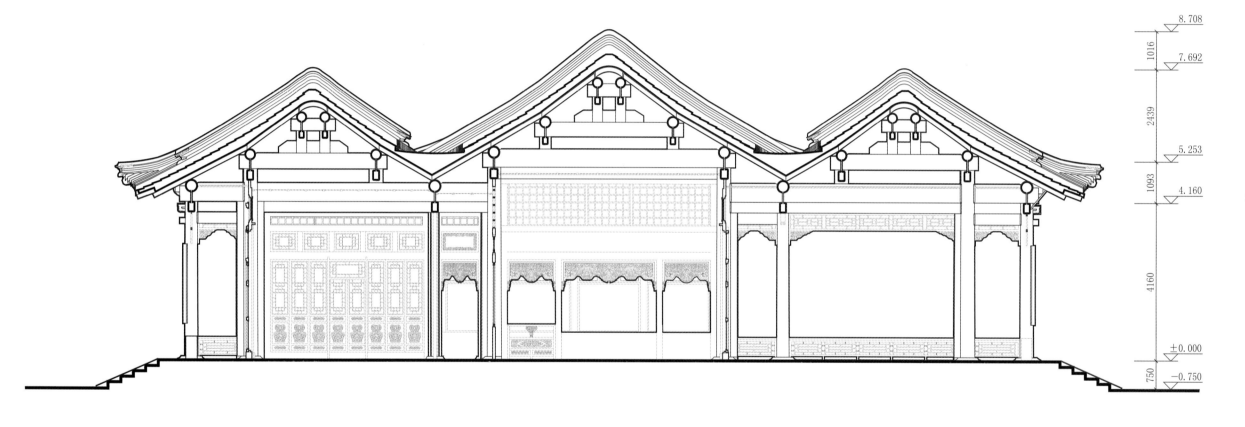

8.708
1016
7.692
2439
5.253
1093
4.160
4160
±0.000
750
−0.750

1610　　5050　　1610　　6300　　1610　　5050　　1610

22840

景福阁明间剖面图

Mingjian Section of the *Jingfu* Pavilion

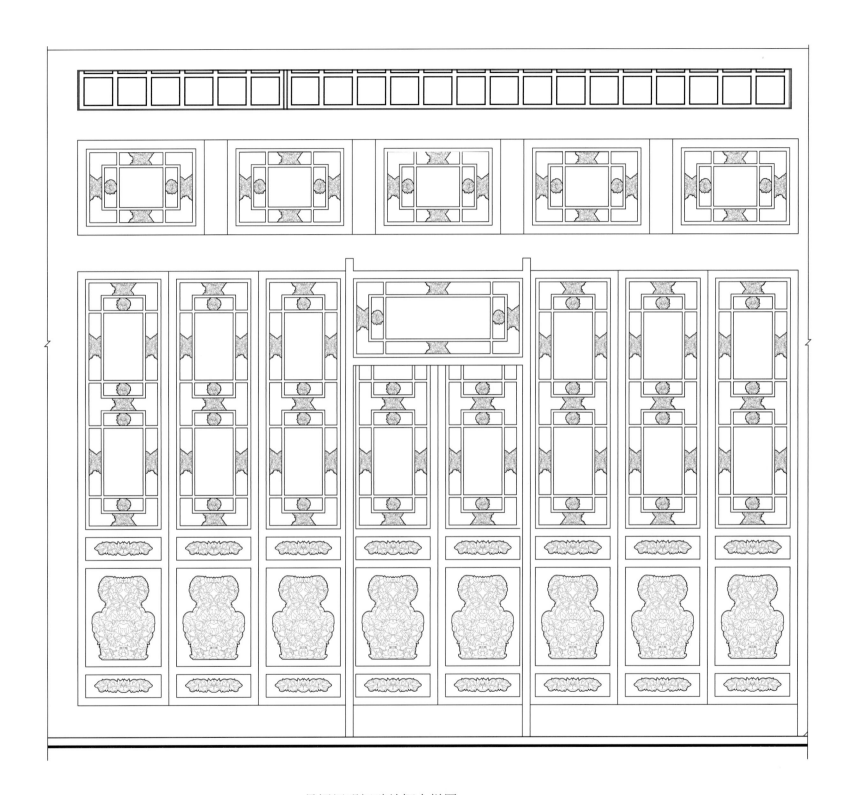

景福阁明间碧纱橱大样图
Bishachu of the Mingjian of *Jingfu* Pavilion

0   0.1   0.5m

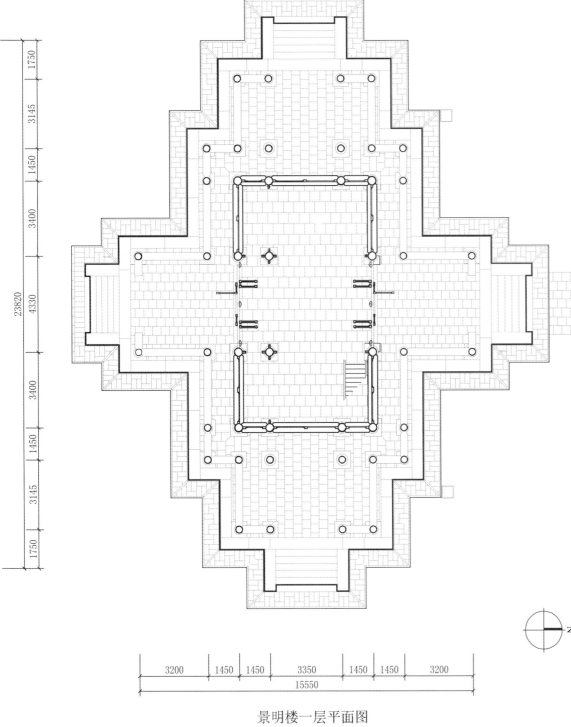

景明楼一层平面图

First Floor Plan of the *Jingming* Pavilion

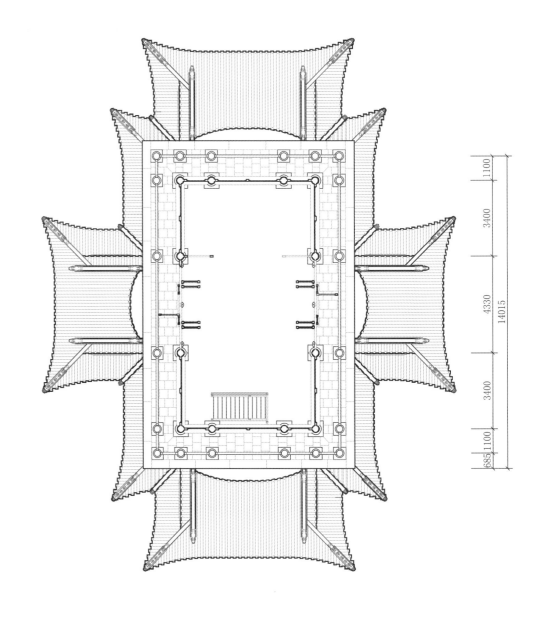

景明楼二层平面图

Second Floor Plan of the *Jingming* Pavilion

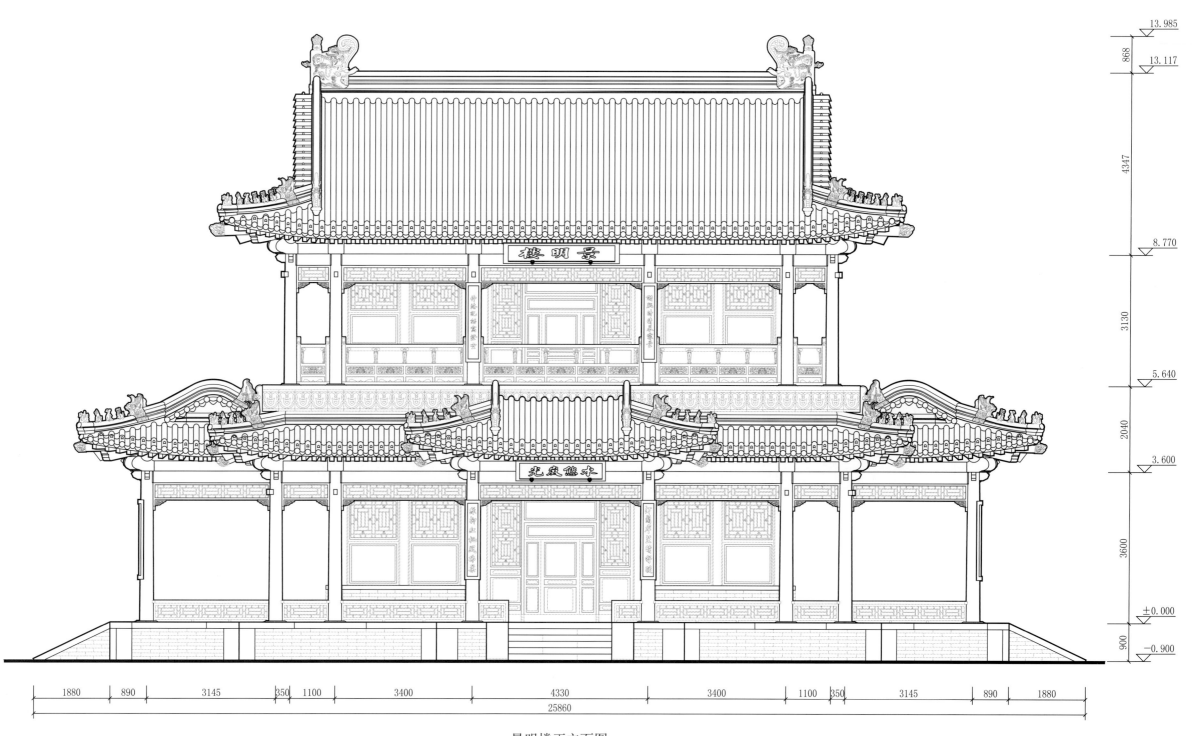

景明楼正立面图

Front Elevation of the *Jingming* Pavilion

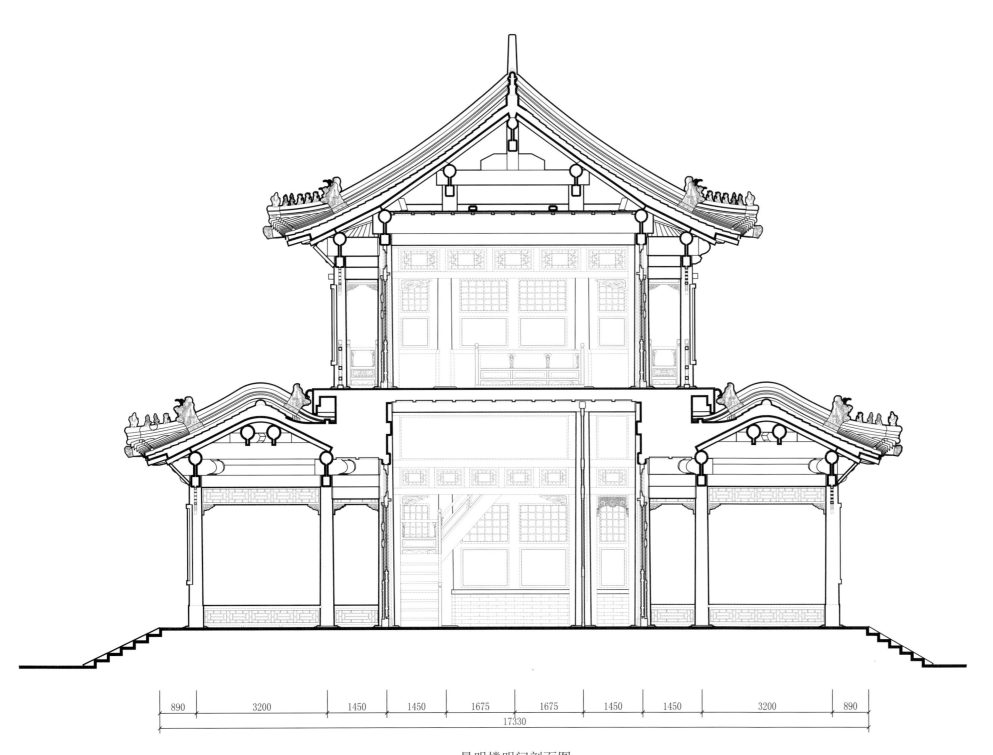

13.985
868
13.117
1362
11.755
2205
9.550
780
8.770
2210
6.560
920
5.640
1700
3.940
620
3.320
3320
±0.000
900
−0.900

890　3200　1450　1450　1675　1675　1450　1450　3200　890
17330

景明楼明间剖面图
Mingjian Section of *Jingming* Pavilion

玉澜堂石座（1980 年代测绘图）
Stone Base of *Yulan* Hall（Drawn in 1980s）

0   10   20cm

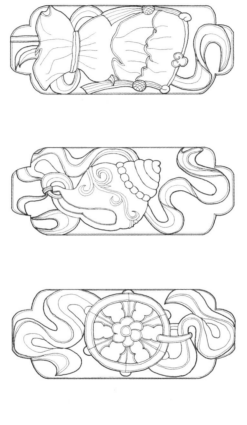

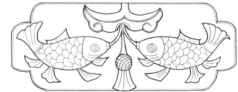

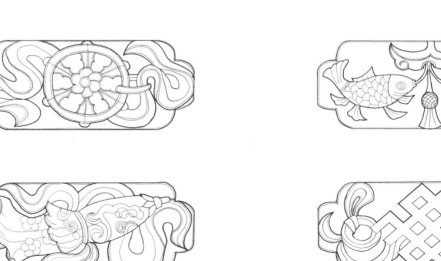

玉澜堂石座（1984 年测绘图）

Stone Base of *Yulan* Hall（Drawn in 1984）

0 　10　　20cm

0 　25　　50cm

乐寿堂前石座（1984 年测绘图）

Stone Base of *Leshou* Hall（Drawn in 1984）

0　　10　　20cm

排云殿石座（1984 年测绘图）
Stone Base of *Paiyun* Hall（Drawn in 1984）

0   5   10   15cm

排云殿石座（1984 年测绘图）

Stone Base of *Paiyun* Hall（Drawn in 1984）

0   5   10   15cm

参与颐和园古建测绘的人员名单

一、1956年、1957年测绘人员

指导教师：卢绳　王淑纯　王龙飞　童鹤龄

1955级学生：

朱良文　董燕军　李鸿博　张秉能　屠兰芬　王嘉勋　李再琛
杨春锦　康玉成　于汗雄　韩蕴琨　史淑莉　崔宝霈　张之清　李祖光
王玲　丁效山

1956级学生：

龙国政　莫剑　生寿春　戚立琇　席绍雯　杨铮　杨月禄
曾秀珍　李鑑荣　王乃香　王作栋　程占文　高士策　石立功　林传良
张家亲　王松澳　冯敬民　黄世强　雷宝乾　李可权　邱曼丽　顾宝和
吴振华　孙国刚　靳士雄　吴暑生　王秩祥　曾玉梅　贾敏章　黄季松
张又泉　任淑贞　范挥寿　郝慎钧　黄尚文　张瑞馨　董有光　赵炳熙
李珍禄　张鼎昕　伍佩瑜　刘姽

二、1979年测绘人员

指导教师：杨道明　高树林　慕春暖

1976级学生：

赵世荣　李良发　王玉华　张建琴　王新华　沈湧麟　杜树山　赵永利
李萍　史继春　王小京

三、1986年测绘人员

指导教师：杨道明　张弛　王兴田　张颀

1984级学生：

韩德重　陆刚　张金辉　袁红　张建蔚　吕毅　董清　潘蓉江
李新　张清　王士杰　韩斌　李甜　齐贵勇　赵东亚　王乐文
蔡珊　金辉　郭彤　骆峰　潘云江　陈小牛　祝捷　张剑平

四、2004—2014年测绘人员

指导教师：王其亨　王蔚　吴葱　曹鹏　丁垚　白成军　张龙

辅导员：王鑫　闫佳亮　来琳　刘丹青
张风梧　梁雪　李哲　陈春红

博士研究生：

2004级：官岿 张慧
2005级：狄雅静
2006级：张龙 张凤梧 殷亮 张宇
2007级：朱蕾 何蓓洁 郭华瞻 赵向东 张宇
2008级：王茹茹
2009级：刘瑜
2010级：伍沙 李婧 袁守愚
2011级：彭飞
2012级：吴晗冰 程枭翀

硕士研究生：

2002级：陈雍
2003级：吴晓冬 张凤梧 阴帅可 张龙 王胜霞 陈春红 张宇 李峥
2004级：郭华瞻 邓宇宁 武菁 李江 陈芬芳 成丽 池小燕 何蓓洁
2005级：陈书砚 宋雪 孔志伟 周文尧 郭俊杰
2006级：李婧 张胜强 王茹茹 何蓉
2007级：马凯 曹苏 曾引 傅东雁 刘瑜 陈学 沙黛诺 陈燕丽 李哲 李琦
2009级：闫金强 李丽娟
2010级：吴晗冰 谭虎 孙倩
2011级：陈双辰 代朋
2012级：雷彤娜 冀凯 周淑玲 徐龙龙

本科生：

2001级：李墨 谭瑶 陆尧 尹慧君 袁琳
2002级：刘瑜 于继成 翟朵朵 何俊乔 刘铭 郝冠民 贺慧 张仲懿
费明喆 丁力 李涵
2003级：付雷 杨梅 高波 富一凝 刘翔宇 周婷 高媛 石崧 杨金莎 伍沙 马睿
刘佳 陈津 袁守愚 刘莹 王庆东 王蕊佳 高冉
徐萌 陈靖 吴南居

2004级：
冯时 陈昱颖 高秀翠 任思为 章强 张伟 戈璧青 贾令德
张韵 李茜（大） 李茜（小） 孙超然 刘峰成 田瑶 沈万夫
张明新 郭志一 王晔楠 高畅 赵颖泽 张备 刘洋
邓晓琳 陈巧如 沈立岑 惠倩楠 何松 刘琨 马婕 陶瑛
王祥 王廷山 辛霖波 杨波 高长宽 耿志鹏 安悦 魏霖霖
王睿 崔孟晓 王雪冬 王悦 孙顿 孙惠玲 王健 郭菁
张磊 耿昀 张瑜 张承 贾明 雷源 于矛 张卫澜 杨帆
翟旭 孙瑜 董辉 李菲 李超 刘哲元 李名岳 陆惠冰
王昊 李世维 高洪波 李治刚 刘利翔 侯影影 陆慧冰 牛晓菲
蒋溢清 沈尧 赵文洲 鲁晓男 周宇黎 马丽丽 张冠兰 姜筱凌
游猎 杨乐 温天伟 邹怡媛 朱英宁 杨盼 庄和峰 常可
王玉 轩煊 阎金强 李琬怡 蒋小敏 刘畅 卢紫荫 何炬

五、此次出版图纸整理人员

图纸审阅：王其亨 丁垚 张龙 张风梧
图纸整理修改：吴晗冰 刘婉琳 耿玥
英文翻译：庄岳 阎晓旭

孙梦 左菲菲 陈铮铮 冯天舒 崔凯 韩楠楠 尹瑾珩 刘思男
胡斌

2005级……
张柯达 陈法琨 黎桂霖 许若木 李晓辉 田可嘉 孙立娜 何熹
孙爱庐 马晓明 郭颖 赵彦超 张溆欣 于天舒 黄兵
李伟平 李洋 张聪恪 张早 王雪培 张骏 陈胜泽
肖聚贤 王浙 李林 王吕双 赵卫松 陈嘉乐
刘倩男 冯海超 李晋 王欢 宁雅静 田健
胡艳娇 王方捷 李振宁 王丹 王天赏 于旋 申林
王剑威 刘家瑞 刘芳 崔强 李茜华 夏 于林
王雅 李波涛 谭旻筠 史磊 杨冬冬 孙德龙
刘倩倩 王洁凝 柳寅生 丁宇 左剑臣 周蕙 孟繁强 吴琳琳
赵鑫甜 张凯敏 孙志九 侯会容 郝鑫 胡晓晨 张坤禹
马炎 粟文倩 李鼎一 詹远 赵轶恬 黄雨尘 郭聪
罗希 董镇彦 姜银辉 朱剡 崔思达 金鑫 陈贞妍
贾文夫 厉庚川 任晓菲 窦晓璐 白泽臣 刘寒 刘程 叶嵩
朱纯瑶 吴昊 范熙矩 郑艳菲 刘文斌 张健 王曦
吴玺 郭青 徐虹 陈晶莹 楚天舒
曹雪 许思扬 张菁 林姗

2008级……
周妙韵 周倩蓉 张园华 黄鹏 侯英春 陈小可 何斌
王典 王子寒 李翔宇 朱琳

2009级……
孟祥健 于莎丽 李安博 陈嗣炫 逄云婷 阮永锦 陶成强 朱曼
张弦 曾良 汪毅 郑翔飞 潘妮 林鹏禹 张子权 卢冬妮
冯小航 林安冬 马凯毅 周玲吉 谢昭沐 胡涵洲 刘苗苗 李姜皓
雷唯 张天娇 李雯婷 江哲麟 邹德华 贾佳 陈永辉
曹梅圳 王顺霞 赵昱 孔振邦 李航 侯新觉 曹哲静 蒋蕊
陈杰 张晓阳 周慧颖 陈霖辉 杨芳灵 刘鑫 张令兮 彭敬怀

2011级……
韩秋爽 侯玉柱 李宗泽 王毅 侯广大 冯彦程 冯胜村 李沁怡
常烁 张李纯一 安秉飞 雷欢玲子 徐玉田 崔雪田 张知
霍丹青 王雪睿 高奇枫 张雪松 刘佩怡 戚一帆 耿玥 崔小瑛
董小雨 邓炎 王雯萱 乔胜星 罗俊杰
王昕宇 吴堃 任宇航 王鑫

Participating Staff in Survey of the Summer Palace

1. Measured&Drawn by (1956 &1957)

Supervisors: LU Sheng, WANG Shuchun, WANG Longfei, TONG Heling

Undergraduate Students:

Class 1955: ZHU Liangwen, DONG Yanjun, LI Hongbo, ZHANG Bingneng, TU Lanfen, WANG Jiaxun, LI Zaichen, ZHANG Wendi, YANG Chunjin, Kang Yucheng, GAN Hanxiong, HAN Yunkun, SHI Shuli, CUI Baopei, ZHANG Zhiqing, LI Zuguang, WANG Ling, DING Xiaoshan

Class1956: LONG Guozheng, MO Jian, SHENG Shouchun, QI Lixiu, XI Shaowen, YANG Zheng, YANG Yuelu, ZHANG Boren, ZENG Xiuzhen, LI Jianrong, WANG Naixiang, WANG Zuodong, CHENG Zhanwen, GAO Shice, SHI Ligong, LIN Chuanliang, ZHANG Jiaqin, WANG Songao, FENG Jingmin, HUANG Shiqiang, LEI Baoqian, LI Kequan, QIU Manli, GU Baohe, WU Zhenhua, SUN Guogang, JIN Shixiong, WU Shusheng, WANG Zhixiang, ZENG Yumei, JIA Minzhang, HUANG Jisong, ZHANG Youquan, REN Shuzhen, FAN Huishou, HAO Shenjun, HUANG Shangwen, ZHANG Ruixin, DONG Youguang, ZHAO Bingxi, LI Zhenlu, ZHANG Dingxin, WU Peiyu, LIU Xian

2. Measured&Drawn by(1979)

Supervisors: YANG Daoming, GAO Shulin, MU Chunnuan

Undergraduate Students:

Class 1976: ZHAO Shirong, LI Liangfa, WANG Yuhua, ZHANG Jianqin, WANG Xinhua, SHEN Yonglin, DU Shushan, ZHAO Yongli, LI Ping, SHI Jichun, WANG Xiaojing

3. Measured&Drawn by(1986)

Supervisors: YANG Daoming, ZHANG Chi, WANG Xingtian, ZHANG Qi

Undergraduate Students:

Class 1984: HAN Dezhong, LU Gang, ZHANG Jinhui, YUAN Hong, ZHANG Jianwei, YU Yi, DONG Qing, PAN Rongjiang, LI Xin, ZHANG Qing, WANG Shijie, HAN Bin, LI Tian, QI Guiyong, ZHAO Dongya, WANG Lewen, CAI Shan, JIN Hui, GUO Tong, LUO Feng, PAN Yunjiang, CHEN Xiaoniu, ZHU Jie, ZHANG Jianping

4. Measured&Drawn by(2004-2014)

Supervisors: WANG Qiheng, WANG Wei, WU Cong, CAO Peng, DING Yao, BAI Chengjun, ZHANG Long, ZHANG Fengwu, LIANG Xue, LI Zhe, CHEN Chunhong

Assistants: WANG Xin, YAN Jialiang, LAI Lin, LIU Danqing

Doctoral Students:

Class 2004: GUAN Wei, ZHANG Hui

Class 2005: DI Yajing

Class 2006: ZHANG Long, ZHANG Fengwu, YIN Liang, ZHANG Yu

Class 2007: ZHU Lei, HE Beijie, GUO Huazhan, ZHAO Xiangdong, CHANG Qinghua

Class 2008: WANG Ruru

Class 2009: LIU Yu

Class 2010: WU Sha, LI JING, YUAN Shouyu

Class 2011: PENG Fei

Class 2012: WU Hanbing, CHENG Xiaochong

Master Students:

Class 2002: CHEN Yong

Class 2003: WU Xiaodong, ZHANG Fengwu, YIN Shuaike, ZHANG Long, WANG Shengxia, CHEN Chunhong, ZHANG Yu, LI Zheng

Class 2004: GUO Huazhan, DENG Yuning, WU Jing, LI Jiang, CHEN Fenfang, CHENG Li, CHI Xiaoyan, HE Beijie

Class 2005: CHEN Shuyan, SONG Xue, KONG Zhiwei, ZHOU Wenyao, GUO Junjie

Class 2006: LI Jing, ZHANG Shengqiang, WANG Ruru, HE Rong

Class 2007: MA Kai, CAO Su, ZENG Yin, FU Dongyan, LIU Yu, CHEN Xue, SHA Dainuo, CHEN Yanli, LI Zhe, LI Qi

Class 2009: YAN Jinqiang, LI Lijuan

Class 2010: WU Hanbing, TAN Hu, SUN Qian

Class 2011: CHEN Shuangchen, DAI Peng

Class 2012: LEI Tongna, JI Kai, ZHOU Shuling, XU Longlong

Undergraduate Students:

Class 2001: LI Mo, TAN Yao, LU Yao, YIN Huijun, YUAN Lin

Class 2002: LIU Yu, YU Jicheng, ZHAI Duoduo, HE Junqiao, LIU Ming, HAO Guanmin, Hehui, ZHANG Zhongyi, FEI Mingzhe, DING Li, LI Han

Class 2003: FU Lei, YANG Chen, GAO Bo, LIU Xiangyu, ZHOU Ting, Gao Yuan, SHI Song, YANG Jinsha, LIU Jia, CHEN Jin, YUAN Shouyu, FU Yining, CHEN Xiao, JIANG Ran, WU Sha, MA Rui, XU Meng, CHEN Jing, WU Nanju, LIU Ying, WANG Qingdong, WANG Ruijia, GAO Ran

Class 2004: FENG Shi, CHEN Yuying, GAO Xiucui, REN Siwei, ZHANG Qiang, ZHANG Wei, GE Biqing, JIA Lingde, ZHANG Yun, LI Qian(elder), LI Qian(younger), SUN Chaoran, LIU Fengcheng, TIAN Yao, SHEN Wanfu, ZHANG Mingxin, GUO Zhiyi, WANG Yangdi, WANG

Yenan, GAO Chang, ZHAO Yingze, ZHANG Bei, LIU Yang, DENG Xiaolin, CHEN Qiaoru, SHEN Licen, HUI Qiannan, HE Song, LIU Kun, MA Jie, Tao Ying, WANG Xiang, WANG Yanshan, XIN Pengfei, Yang Bo, GAO Changkuan, GENG Zhipeng, AN Yue, WEI Linlin, WANG Rui, CUI Mengxiao, WANG Xuedong, WANG Yue, SUN Di, SUN Huiling, WANG Jian, GUO Jing, ZHANG Lei, GENG Yun, DONG Hui, LI Fei, LI Chao, LIU Zheyuan, LI Mingyue, LU Huibing, ZHAI Xu, SUN Yu, ZHANG Cheng, JIA Ming, LEI Yuan, YU Mao, ZHANG Weilan, YANG Fan, WANG Hao, LI Shiwei, GAO Hongbo, LI Zhigang, LIU Lixiang, HOU Yingying, LU Huibing, NIU Xiaofei, JIANG Yiqing, SHEN Yao, ZHAO Wenzhou, LU Xiaonan, ZHOU Yuli, MA Lili, ZHANG Guanlan, JIANG Xiaoling, YOU Lie, YANG Le, WEN Tianwei, ZOU Yiyuan, ZHU Yingning, YANG Pan, ZHUANG Hefeng, Chang Ke, WANG Yu, XUAN Xuan, YAN Jinqiang, LI Wanyi, JIANG Xiaomin, LIU Chang, LU Ziyin, HE Ju, SUN Meng, ZUO Feifei, CHEN Zhengzheng, FENG Tianshu, CUI Kai, HAN Nannan, YIN Jinheng, LIU Sinan, HU Bin

Class 2005: ZHANG Keda, CHEN Fakun, LI Guilin, XU Ruomu, LI Xiaohui, TIAN Kejia, SUN Lina, HE Xi, SUN Ailu, MA Xiaoming, GUO Ying, ZHAO Yanchao, ZHANG Xuxin, YU Tianshu, TAO Sui, HUANG Bing, LI Weiping, LI Yang, ZHANG Congke, ZHANG Zao, WANG Xuepei, ZHAN Yang, ZHANG Jun, CHEN Shengze, XIAO Juxian, WANG Xi, FENG Haichao, LI Lin, WANG Jin, LYU Shuang, ZHAO Weisong, CHEN Jiale, LIU Qiannan, WANG Fangjie, LI Zhenning, WANG Dan, WANG Huan, WANG Tianshang, NING Yajing, TIAN Jian, HU Yanjiao, LIU Jiarui, LIU Fang, CUI Qiang, LI Qian, HUAN Xia, YU Xuan, SHEN Lin, WANG Jianwei, LI Ya, LI Botao, LIU Yinsheng, TAN Minyun, SHI Lei, YANG Dongdong, SUN Delong, ZHAO Xintian, LIU Qianqian, WANG Jiening, DING Yu, ZUO Jianbing, GUO Jiasheng, HU Xiaochen, ZHANG Kunyu, MA Yan, SU Wenqian, ZHANG Kaimin, SUN Zhijiu, HOU Huirong, HAO Xin, ZHOU Qianbin, XU Longlong, LUO Xi, DONG Zhenyan, JIANG Yinhui, LI Dingyi, ZHAN Yuan, ZHAO Yitian, HUANG Yuchen, GUO Cong, JIA Wenfu, LI Na, REN Xiaofei, TUO Chuan, ZHU Yan, CUI Sida, JIN Xin, CHEN Zhenyan, ZHU Chunyao, WU Hao, FAN Xixuan, DOU Xiaolu, BAI Zechen, ZHOU Hui, MENG Fanqiang, WU Linlin, WU Xi, GUO Qing, XU Hong, ZHENG Yanfei, LIU Wenbin, Liu Qian, LIU Cheng, Ye Song, CAO Xue, XU Siyang, ZHANG Jing, LIN Shan, CHEN Jingying, CHUN Tianshu, ZHANG Jian, WANG Xi, ZHOU Miaoyun, ZHOU Qianrong, ZHANG Yuanhua, HUANG Peng, HOU Yingchun, CHEN Xiaoke, HE Bin

Class 2008: WANG Dian, WANG Zihan, LI Xiangyu, ZHU Lin

Class 2009: MENG Xiangjian, YU Lisha, LI Anbo, CHEN Sixuan, PANG Yunting, RUAN Yongjin, TAO Chengqiang, ZHU Man, ZHANG Xian, ZENG Liang, WANG Yi, ZHENG Xiangfei, PAN Ni, LIN Pengyu, ZHANG Ziquan, LU Dongni, FENG Xiaohang, LIN Andong, MA Kaiyi, ZHOU Lingji, XIE Zhaomu, HU Hanzhou, LIU Miaomiao, LI Jianghao, LEI Wei, ZHANG Tianjiao, LI Wenting, JIANG Zhelin, ZOU Dehua, DU Songyi, JIA Jia, CHEN Yonghui, CAO Meizhen, WANG Shunxia, ZHAO Yu, KONG Zhenbang, LI Hang, HOU Xinjue, CAO Zhejing, JIANG Rui, CHEN Jie, ZHANG Xiaoyang, ZHOU Huiying, CHEN Linhui, YANG Fangling, LIU Xin, ZHANG Lingxi, PENG Jinghuai, WANG Xinyu, WU Kun, REN Yuhang, WANG Xin

Class 2011: HAN Qiushuang, Hou Yuzhu, LI Zongze, WANG Yi, HOU Guangda, FENG Yancheng, FENG Shengcun, LI Qinyi, CHANG Shuo, ZHANG Lichunyi, AN Bingfei, LEI Huanlingzi, XU Yutian, CUI Xue, ZHANG Zhi, HUO Danqing, WANG Xuerui, GAO Qifeng, ZHANG Xuesong, LIU Peiyi, QI Yifan, Geng Yue, CUI Xiaoying, DONG Xiaoyu, DENG Yan, WANG Wenxuan, QIAO Shengxing, LUO Junjie

5. Drawings Arrangement

Check and Proof: WANG Qiheng, DING Yao, ZHANG Long, ZHANG Fengwu

Collation and Revision: WU Hanbing, LIU Wanlin, Geng Yue

English Translation: ZHUANG Yue, YAN Xiaoxu

图书在版编目（CIP）数据

颐和园 = SUMMER PALACE（2nd Edition）: 汉、英 /
王其亨主编；张龙，张凤梧编著. — 2 版 . — 北京：
中国建筑工业出版社，2023.12
（中国古建筑测绘大系 . 园林建筑）
ISBN 978-7-112-28852-6

Ⅰ. ①颐… Ⅱ. ①王… ②张… ③张… Ⅲ. ①颐和园
－古建筑－图集 Ⅳ. ① TU-87

中国国家版本馆 CIP 数据核字（2023）第 163622 号

丛书策划 / 王莉慧
责任编辑 / 柳　冉　李　鸽
书籍设计 / 付金红
责任校对 / 王　烨

中国古建筑测绘大系·园林建筑

颐和园（第二版）

天津大学建筑学院
北京市颐和园管理处　编写

王其亨　主编　张龙　张凤梧　编著

Traditional Chinese Architecture Surveying and Mapping Series:
Garden Architecture
SUMMER PALACE（2nd Edition）
Compiled by School of Architecture,Tianjin University &
Administration Office of the Summer Palace,Beijing
Chief Edited by WANG Qiheng
Edited by ZHANG Long,ZHANG Fengwu

*

中国建筑工业出版社出版、发行（北京海淀三里河路 9 号）

各地新华书店、建筑书店经销

北京方舟正佳图文设计有限公司制版

北京雅昌艺术印刷有限公司印刷

*

开本：787 毫米 ×1092 毫米　横 1/8　印张：36　字数：926 千字
2023 年 12 月第二版　2023 年 12 月第一次印刷
定价：**288.00** 元
ISBN 978-7-112-28852-6
（41267）